Advanced Chromatic Monitoring

Series in Sensors

Series Editors: Barry Jones and Haiying Huang

Other recent books in the series:

For more information about this series, please visit: https://www.crcpress.com/physics/sensor-science-technology

Advanced Chromatic Monitoring

Edited by
Gordon R. Jones
Joseph W. Spencer

CRC Press
Taylor & Francis Group
Boca Raton London New York

CRC Press is an imprint of the
Taylor & Francis Group, an **informa** business

First edition published 2021
by CRC Press
6000 Broken Sound Parkway NW, Suite 300, Boca Raton, FL 33487-2742

and by CRC Press
2 Park Square, Milton Park, Abingdon, Oxon, OX14 4RN

Library of Congress Cataloging-in-Publication Data
Names: Jones, G. R. (Gordon Rees), 1938- editor. | Spencer, Joseph W., editor.
Title: Advanced chromatic monitoring / edited by Gordon R. Jones, Joseph W. Spencer.
Description: First edition. | Boca Raton : CRC Press, 2020. | Series:
    Series in sensors | Includes bibliographical references and index.
Identifiers: LCCN 2020017220 | ISBN 9780367409470 (hardback) | ISBN 9780367815202 (ebook)
Subjects: LCSH: Quantum chromodynamics. | Chromatographic analysis. | Computational complexity.
Classification: LCC QC793.3.Q35 A38 2020 | DDC 543/.8--dc23
LC record available at https://lccn.loc.gov/2020017220

ISBN: 978-0-367-40947-0 (hbk)
ISBN: 978-0-367-81520-2 (ebk)

Typeset in Times LT Std
by Nova Techset Private Limited, Bengaluru & Chennai, India

Visit the Taylor & Francis Web site at
http://www.taylorandfrancis.com

and the CRC Press Web site at
http://www.crcpress.com

Contents

SECTION I Basic Chromatic Principles

SECTION II Chromatic Monitoring of Liquids and Biological Tissue

SECTION III Chromatic Monitoring of Mechanical Vibrations

SECTION IV   Environmental Applications

SECTION V   Advanced Chromatic Monitoring

Acknowledgements

The interaction with several organisations over several years has enabled chromatic technology to be advanced, with many examples which are presented in this book. The contributions of the following organisations are much appreciated:

AREMA, USA
Electricity North West Ltd.
Fairbanks Environmental Ltd.
Finch Electronics Ltd.
Fraunhofer Institute, Germany
National Grid
National Oceanographic Centre
Northern Powergrid
Renewable Energy Foundation, UK
Sensor City Liverpool Ltd.
TNB Research Sdn. Bhd., Malaysia
Western Power Distribution

The assistance and support provided by Dr. D. H. Smith is particularly appreciated, not only for his technical input on many projects described in this book but also for his efforts in finalising to specification the figures which appear in this book.

Contributors

Ziyad S. D. Almajali
Faculty of Engineering
Mutah University
Mutah, Jordan

A. A. Al-Temeemy
Department of Laser and Optoelectronics
    Engineering
Al-Nahrain University
Baghdad, Iraq

E. Elzagzoug
Manufacturing Intelligence Group
Advanced Manufacturing Research Centre
    (AMRC Factory 2050)
Sheffield, United Kingdom

C. Garza
British Telecom (International Delivery)
Madrid, Spain

G. R. Jones
Department of Electrical Engineering and
    Electronics
University of Liverpool
Liverpool, United Kingdom

J. L. Kenny
Brimstage Engineering Solutions Ltd.
Merseyside, United Kingdom

J. Lawton
Department of Electrical Engineering and
    Electronics
University of Liverpool
Liverpool, United Kingdom

M. Ragaa
Project Delivery Engineer
Northern Powergrid
Wakefield, United Kingdom

H. M. Shabeer
Head of Chromatic Mobile Health
    Technologies Private Limited
Thengana Medical Mission Hospital and
    Research Centre
Perumpanachy P.O. Changanacherry
Kottayan, Kerala, India

L. M. Shpanin
Department of Engineering and
    Mathematics
Sheffield Hallam University
Sheffield, United Kingdom

D. H. Smith
Department of Electrical Engineering and
    Electronics
University of Liverpool
Liverpool, United Kingdom

L. U. Sneddon
Director of Bioveterinary Science
University of Liverpool
Liverpool, United Kingdom

J. W. Spencer
Department of Electrical Engineering and
    Electronics
University of Liverpool
Liverpool, United Kingdom

A. T. Sufian
Liverpool John Moores University
Liverpool, United Kingdom

R. K. Todd
Remote Services Technology
Chester, Cheshire

Z. Wang
R&D Senior Engineer
ABB (China) Limited
Chaoyang, Beijing, China

J. D. Yan
Department of Electrical Engineering and
   Electronics
University of Liverpool
Liverpool, United Kingdom

Prologue

An overview is provided of how basic chromatic principles (Jones, G. R. et al., 2008) have been further developed into new domains, along with practical examples of their deployment. The emphasis is on the practicalities of the method for diverse applications.

As technology in general advances, there are extensive ways in which it may be deployed for measurement and to produce an extensive amount of data. The acquisition of such large amounts of data often contains much unnecessary data which effectively constitutes a high level of noise and so masks the required data whilst inflating cost. Many approaches are available which seek to address the problem in different ways – for example, through the use of artificial intelligence.

The chromatic approach may be regarded as addressing this problem from a different perspective. It recognises the undesirable need for extra measurement capability alone and recognises the fact that information comes from monitoring, which should not be confused with measurement but is the means whereby diagnosis is made. The approach enables emerging conditions to be recognised without the need for an excessive amount of measured data.

The present book shows that there are different ways of deploying chromaticity for particular applications. It provides examples which can form the basis of yet more future developments.

It illustrates how the use of chromatic monitoring has evolved from only the use of optical fibre-based sensing (Jones, G. R. et al., 2008) through the use of PC-based systems using VDU screen illumination as the optical source and a miniature camera for detection, leading to the production of a self-contained, controlled unit for use at remote sites without the need for a PC (Chapter 2).

Advances have been made in the domain of monitoring mechanical vibrations (Chapter 3) and in a variety of environmental monitoring, ranging from marine water monitoring and power from wind production to assessing the environmental impact of various electrical insulation gases (Chapter 4). Advances have also been made in adapting chromaticity for multidimensional monitoring such as elderly care and fish monitoring, three-phase power-line monitoring and time–frequency and time–wavelength monitoring, as well as a combination of different factors for assessing the degradation of high-voltage transformer oils and different gases for high current interruption.

The techniques described have the potential for deployment in several other applications in a cost-effective and convenient-to-use manner (e.g. mobile phone neonate in vivo jaundice monitoring).

Section I

Basic Chromatic Principles

A brief summary of the basic concepts of the chromatic monitoring approach is presented which has enabled chromatic monitoring concepts to be further developed, leading to their deployment for a variety of different applications, ranging from preliminary medical monitoring via environmental issues to electric power system monitoring.

1 Overview of Chromatic Monitoring

G. R. Jones and J. W. Spencer

CONTENTS

1.1 INTRODUCTION

This chapter summarises the concept of chromaticity and its basic attributes for advancing its capabilities beyond those already established (Jones et al., 2008a). It indicates how chromaticity is related to pure measurement and the diagnosis of a condition. It explains how chromaticity quantifies information in terms of signal properties via an optimum of only three parameters. Approaches for extracting values of these three parameters via chromatic processors are described, and the manner in which these values may be displayed via chromatic maps and calibration graphs is indicated. The concepts of primary and secondary chromatic monitoring are explained, and the manner in which continuous and discrete signals can be addressed to check for emerging information about complex conditions is indicated. These aspects form the basis from which more recent advances have been made.

1.2 MONITORING AND CHROMATICITY

Monitoring may be regarded as assessing the condition of a system or component (Jones et al., 2008a). It can be regarded as a means for connecting *measurement* with *diagnosis* (Figure 1.1). *Measurement* (which is reduction in nature) provides accurate quantification, whereas *diagnosis* involves distinguishing between conditions. Thus, *monitoring* provides a means for interconnecting *measurement* and *diagnosis* and dealing with *complexity* (Jones et al., 2008a).

One characteristic of complexity is *emergence*, that is, dealing with unexpected events. *Chromaticity* addresses *emergence* via cross-correlation with processors whose responses overlap (non-orthogonal) rather than being discrete. An example of such non-orthogonal processors (Jones et al., 2008a) is shown in Figure 1.2 with three processors (R, G, B) whose responses overlap in the measurement domain. The distribution of a signal (U) addressed by the three R, G, B processors is also shown in Figure 1.2.

An example of such monitoring is the capability of the human vision system to distinguish between different optical colours (e.g., Billmeyer and Saltzman, 1981) with only three non-orthogonally

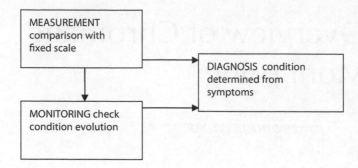

FIGURE 1.1   Monitoring in relation to measurement and diagnosis.

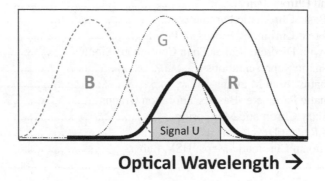

Optical Wavelength →

FIGURE 1.2   Three non-orthogonal processors (R, G, B) superimposed upon a signal (U).

responsive processors (R, G, B), where R addresses the longer optical wavelengths, G the middle wavelengths and B the shorter wavelengths.

1.3   EXAMPLES OF BASIC CHROMATIC PARAMETERS

The outputs of the three non-orthogonal processors (Ro, Go, Bo) may be treated mathematically to produce various chromatic parameters, as shown by Jones et al. (2008a). Details of two particular examples, which have been shown to be useful for widespread monitoring applications, are shown in Table 1.1. These parameters quantify various features of a signal. One set of such parameters, which can be calculated from Ro, Go, Bo, are H, L, S, defined by the mathematical formulae given in Table 1.1.

L is the strength of the signal, which is the effective area under the signal envelope (Figure 1.3). H is the hue, which indicates the dominant part of the signal. S is the saturation, which quantifies the difference between the maximum and minimum levels of the R, G, B components of the signal (and so also indicates the spread of the signal).

An alternative set of chromatic parameters that have been shown to be useful for monitoring applications (Jones et al., 2008a) are X, Y, Z. These three parameters represent the relative magnitudes of the outputs from each processor, Ro, Go, Bo, as indicated in Table 1.1. As such, they provide an approximate indication of the signal distribution.

These two sets of chromatic parameters may be used to form chromatic maps on which a particular signal may be represented. Investigations (Jones et al., 2008a) have shown that a particularly useful map for monitoring applications is a polar diagram of L versus H (Figure 1.4a). This provides an indication of signal strength (L) as the radial coordinate and dominant range of the signal (H) as the azimuthal coordinate. More details of the signal R, G, B structure are provided by a Cartesian map

TABLE 1.1

Mathematical Definitions of Some Conventional Chromatic Parameters (Jones et al., 2008a)

Parameter	Expression	Name	Physical Meaning
HLS			
L	(Ro + Go + Bo)/3	Strength	Signal Area
H		Hue	Dominant Part
	120 − 120r/(g + r) b = min		r = Ro-min(Ro, Go, Bo)
	240 − 120g/(g + b) r = min		g = Go-min(Ro, Go, Bo)
	360 − 120b/(b = r) g = min		b = Bo-min(Ro, Go, Bo)
S	[(m+) − (m−)] / [(m+) + (m−)]	Saturation	Spread
			(m+) = max(Ro,Go,Bo)
			(m−) = min(Ro,Go,Bo)
XYZ:			
X	Ro/3L		R Proportion
Y	Go/3L		G Proportion
Z	Bo/3L		B Proportion

X + Y + Z = 1; X = Y = Z = 0.33; X = 1 − Z when Y = 0.
S = 0 → Uniform Signal; S = 1 → Monochromatic.

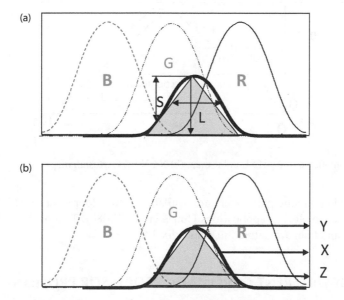

FIGURE 1.3 Physical meaning of various chromatic parameters (a) signal strength (L), hue (H) and saturation (S). (b) relative magnitudes X, Y, Z.

of X versus Z, which provides an indication of the relative magnitudes of the R, G, B components and hence distribution of the signal (Figure 1.4b). (It should be noted that the XYZ chromatic map incorporates values of X as well as Y and Z by virtue of X + Y + Z=1.) Taken together, these two maps provide a quantification of the signal, distinguishing features of signal strength, dominant region and proportion of signal strength spread over the R, G, B zones.

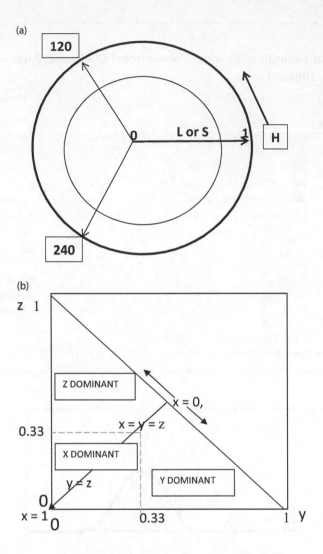

FIGURE 1.4 Chromatic maps (a) H : L polar map. (b) X : Y : Z Cartesian map.

Some additional chromatic parameters are available from colour science which can be adapted for some particular monitoring applications (Jones et al., 2008b). Two of these additional sets of parameters (HSV and Lab) are defined in Appendix 1A.

1.4   EFFECT OF DIFFERENT CHROMATIC PROCESSOR PROFILES

The extent and shape of the chromatic R, G, B profiles may be varied to provide the most appropriate information extraction (Jones et al., 2008a) for particular applications. Figure 1.5a (i) shows three R, G, B Gaussian profiles overlapping at their midpoints. With such processor profiles, the value of the dominant range (H) for a monochromatic signal displaced along the signal axis is shown in Figure 1.5a (ii). The resultant H distribution shows some non-linear variation, thus leading to a moderately variable sensitivity.

Figure 1.5b (i) shows three R, G, B Gaussian profiles with only minimal overlaps, whilst Figure 1.5b (ii) shows the corresponding H variation for a monochromatic signal shifted along the signal axis. This has a highly non-linear variation of H, implying a highly non-linear sensitivity.

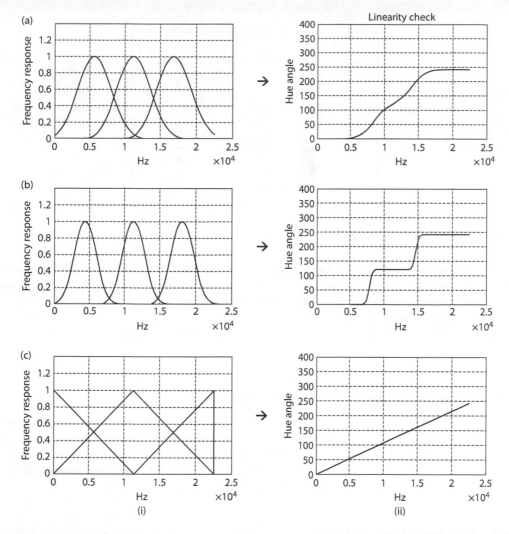

FIGURE 1.5 Effect of R, G, B profile shape and overlap on processing. (a) Gaussian profiles, midpoint overlaps (i) profiles (ii) sensitivity. (b) Gaussian profiles of R, G, B, minimal overlaps (i) profiles (ii) sensitivity. (c) Linear profiles, midpoint overlaps (i) profiles (ii) sensitivity. (Jones, G.R. et al. (2008a) *Chromatic Monitoring of Complex Conditions*. CRC Press, ISBN 978-1-58488-988-5, figures 2.2.2 (a), (b), (c) and 2.2.3 (a), (b), (c) combined.)

Figure 1.5c (i) shows three R, G, B processors with linear variation profiles overlapping at their midpoints. Figure 1.5c (ii) shows a highly linear variation of H for this case so that the sensitivity is relatively constant throughout the overall range.

An appreciation of these properties is important for the further development of chromatic monitoring.

1.5 CHROMATICITY WITH DISCRETE RATHER THAN CONTINUOUS SIGNALS

Discrete signal processing involves addressing a series of separate signals (rather than a single continuous one) with three non-orthogonal processors (Jones et al., 2008a). Such discrete signals may correspond to outputs from a number of different sensors operating in parallel (Figure 1.6). The signals may be divided into three groups, each group corresponding to different measured parameters (e.g., set 1 being a series of temperatures and corresponding to R, set 2 being a series

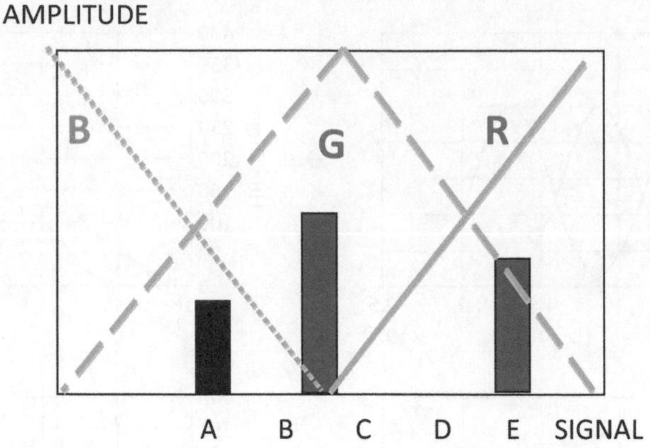

AMPLITUDE

FIGURE 1.6   Discrete signals addressed by three non-orthogonal processors.

of electric currents and corresponding to G, set 3 being a series of gas pressures and corresponding to B). Each group is addressed by one of three non-orthogonal processors (R, G, B). The chromatic parameters derived from the outputs of R, G, B provide a comparison of the various sensor outputs so that the relative magnitudes of the groups can be defined. An example of such a system is the monitoring of different gases dissolved in transformer oils, which can indicate different causes of the oil degradation (Jones et al., 2008a).

1.6   VARIOUS CHROMATIC REPRESENTATIONS

Consideration of previous research (Jones et al., 2008a) shows that chromatically monitored data may be represented in different ways in order to convey information in the most appropriate manner for different purposes. Three main ways in which this may be achieved are via chromatic trend mapping, chromatic parameter-based calibration graphs and secondary chromaticity. These provide the basis of future developments and applications of chromatic monitoring.

1.6.1   Chromatic Trend Mapping

The H : L and X, Y, Z chromatic maps (Figure 1.4a and b) provide a convenient means for classifying monitored signals. A test result shown on such maps indicates the condition of the monitored system according to the values of the chromatic map coordinates (H, L or X, Y, Z). Variation in the condition of the system is indicated via the locus of such measured coordinate values. Figure 1.7a shows an example of such a variation on an XYZ Cartesian chromatic map, whilst Figure 1.7b shows a variation on an H : L polar chromatic map. This illustrates the power of such chromatic mapping for providing a rapid and convenient impression of trends in a complex system.

   As a result, the H : L map provides an indication of how the strength (L) and dominant value (H) vary, whilst the X, Y, Z map gives an indication of the approximate distribution of the relative magnitudes of three of the signal's components.

1.6.2   Chromatic Parameter-Based Calibration Graphs

Many applications require a quantification of an effect in order to indicate how far from a critical level the condition may be. A chromatic parameter can be derived to provide such quantification by calibration with known conditions. A first step in the choice of the most appropriate chromatic parameter is checking for a required trend on the XYZ and H : L chromatic maps [Figure 1.7a (i)

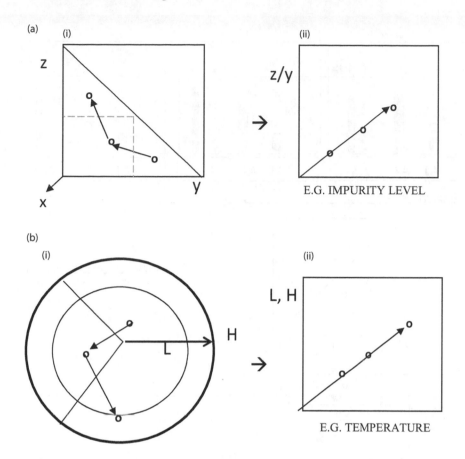

FIGURE 1.7 Representation of system trends on chromatic maps (i) and calibration graphs (ii). (a) X, Y, Z chromatic Cartesian map and Z/Y calibration graph. (b) H : L chromatic polar map and L or H calibration graph.

and b (i)]. A more precise calibration curve may then be produced by more detailed monitoring tests with regard to that chromatic parameter.

Examples of such chromatic parameter calibration curves are shown in Figure 1.7a (ii) and b (ii) with the chromatic parameters Z/Y as a function of impurity level [Figure 1.7a (ii)] and H or L [Figure 1.7b (ii)] as a function of temperature.

Choices may be made based upon which chromatic parameter might provide the best trend. For example, consideration may be given to which parameter shows the most linear variation over the range of the monitored physical condition. Alternatively, the parameter providing the highest sensitivity within a given range may be preferred. Furthermore, a combination of chromatic parameters may be preferred, for example, Z/Y=B/G.

1.6.3 SECONDARY CHROMATIC MONITORING

There are many cases where the time variation of a complex condition may need to be observed and quantified, for example, degradation of a liquid with age (Elzagzoug et al., 2014). In such a case, the condition may be chromatically tracked at selected intervals of time (t1, t2, t3), as shown in Figure 1.8a. This is referred to as *primary chromatic monitoring*. A suitable chromatic parameter from primary chromatic monitoring is then chosen and tracked as a function of time. *Secondary chromatic monitoring* may then be undertaken in the time domain using this parameter and secondary chromatic maps produced. Various trends may then be observed in the resulting secondary chromatic maps (Figure 1.8b).

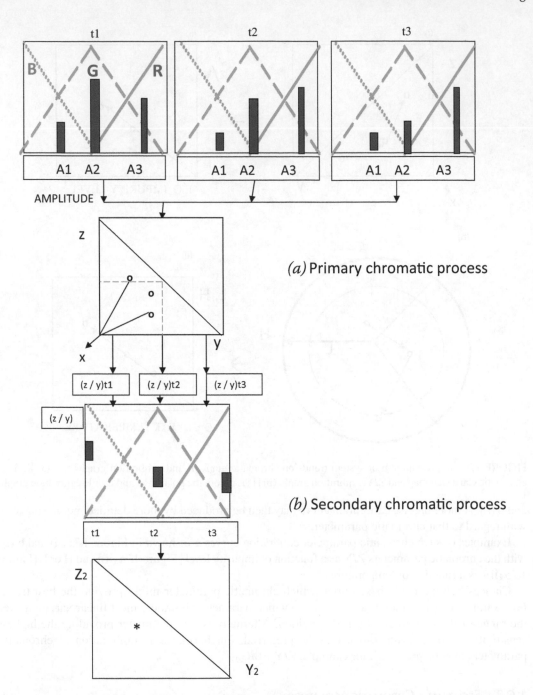

FIGURE 1.8   Secondary chromatic processing. (a) Primary processing step. (b) Secondary processing step.

1.7   SUMMARY AND RECENT DEVELOPMENTS

Chromatic techniques have already been applied for monitoring a variety of complex conditions, including electric power equipment, medical diagnosis and environmental aspects (Jones et al., 2008a). From these studies, it is apparent that chromatic techniques have potential for further monitoring developments. The present book describes progress made in such extrapolation of chromatic monitoring.

Details of the chromatic methodologies described in this chapter are based upon preferences demonstrated by previously used investigations (Jones et al., 2008a). Consideration has mainly been made of the widely used chromatic parameters HLS and XYZ plus their properties and chromatic maps. The significance of such maps has been indicated for providing a rapid and convenient impression of trends in a complex system. Such maps can also be used for identifying an appropriate chromatic parameter for use as a calibration factor in the more detailed quantification of the variation of a physical condition. Secondary chromatic monitoring has been indicated to have the potential to provide a further level of chromatic analysis, such as the time variation, of a different domain chromatic condition. Discrete chromatic processing has been used for monitoring an array of different signals.

Further developments of chromatic monitoring have occurred for many areas of application, ranging from the optical domain (e.g., combining different optical properties) via the acoustical domain to the processing of a multiplicity of complex discrete components. The chromatic monitoring approach has been extended for use with different instrumentation and devices in the quest for improved efficiency plus convenient and economic monitoring. For example, such developments have involved the use of mobile phone cameras and visual display units as optical sources and devices for pre-data transmission processing. These various examples are considered in the chapters which follow.

APPENDIX 1A: ADDITIONAL CHROMATIC MAPS: HSV, LAB

There are transformations other than the HLS and XYZ which are well established in colour science and which can have potential for some particular chromatic monitoring applications (Jones et al., 2008b). Two such transformations are the H, S, V and L, a, b methods. The mathematical formulations of these transformations are presented in Table 1A.1 for convenient comparison with the H, L, S and X, Y, Z transformations given in Table 1.1.

TABLE 1A.1

Mathematical Definitions of Some Additional Colour Science Parameters (Rogers, 1985; Ainouz et al., 2006)

Parameter	Expression	Physical Meaning
HSV		
V	$\max(R, G, B)$	Highest output
S	$(V - \min)/V$	Relative spread of V
		$\min = \min(R, G, B)$
H	$cb - cg$	V = R dominant
	$2 + cr - cb$	V = G dominant
	$4 + cg - cr$	V = B dominant
		$cr = (V - R)/(V - \min)$
		$cg = (V - G)/(V - \min)$
		$cb = (V - B)/(V - \min)$
		$H = H*60$
		$H < 0 \ H = H + 360$
Lab		
L	$116(G/Gn)*1/3 - 16$	Normalised mid processor
a	$500[(R/Rn)*1/3 - (G/Gn)*1/3]$	Long → mid difference
b	$200[(G/Gn)*1/3 - (B/Bn)*1/3]$	Mid → short difference
		Rn, Gn, Bn references
		For R/Rn, G/Gn, B/Bn > 0.008856

REFERENCES

Ainouz, S., Zallet, J., de Martino, A. and Colledt, C. (2006) Physical interpretation of polarisation-encoded images by colour preview. *Opt. Express* 14(13), 5916–5927.

Billmeyer, F.W. and Saltzman, M. (1981) *Principles of Color Technology*. John Wiley, New York.

Elzagzoug, E., Jones, G.R., Deakin, A.G. and Spencer, J.W. (2014) Condition monitoring of high voltage transformer oils using optical chromaticity. *Meas. Sci. Technol* 25, 065205.

Jones, G.R., Deakin A.G. and, Spencer, J.W. (2008a) *Chromatic Monitoring of Complex Conditions*. CRC Press, ISBN 978-1-58488-988-5.

Jones, G.R., Pavlova, P. and Spencer, J.W. (2008b) *Other Chromatic Processing Algorithms, Chapter 3, Chromatic Monitoring of Complex Systems*. CRC Press, ISBN 978-1-58488-988-5.

Rogers, D. (1985) *Procedural Elements for Computer Graphic*. McGraw-Hill, New York.

Section II

Chromatic Monitoring of Liquids and Biological Tissue

The evolution of chromatic monitoring for liquid and biological tissue monitoring is described. It shows how chromatic techniques evolved from a basic optical fibre system for monitoring petroleum fuels, to a personal computer (PC)–based desktop system using visual display unit (VDU) illumination for monitoring *E. coli* in urine, followed by a portable PC system for honey monitoring in remote geographical regions, to an adaptation based upon the use of a mobile phone for *in vivo* neonate jaundice monitoring, to the production of a compact, portable unit for transformer oil monitoring with remote wireless addressing.

2 General Overview of Liquid Chromatic Monitoring

G. R. Jones, A. T. Sufian and D. H. Smith

CONTENTS

2.1 INTRODUCTION

Chromatic optical monitoring of various conditions may be regarded as having focussed upon the use of two classes of systems – optical fibre and camera systems. Both systems have broadly similar structures (Figure 2.1), each of which involves an illumination source whose output is directed onto a sensor element where the light is modulated before transmission to a chromatic optical detector. The detector output is processed to produce chromatic parameters (e.g., H, L, S; Chapter 1) which indicate the condition of the sample. Examples of optical fibre and camera monitoring units are shown in Figure 2.2a and b, which have been deployed for monitoring the condition of the liquid digestate at a biodigestion processing site Rallis et al. (2005). This indicates the different arrangements with each system – the optical fibre system has an optical probe which is immersed into the liquid, whilst the camera system is nonintrusive and addresses the liquid remotely.

A list of some parameters which have been addressed chromatically by optical fibre and camera systems is given in Figure 2.3. This indicates that both optical fibre and camera systems are capable of monitoring various liquids, the choice depending upon various factors such as access to a liquid/tissue sample, portability or being site based, effective cost, flexibility and so on.

2.2 CAMERA-BASED SYSTEMS OPTIONS

The camera-based system offers much flexibility in being adapted for various purposes and also in the capability of its operation to be automatically varied even during its use. As camera-based technology continues to evolve, the use of such systems for chromatic monitoring is also evolving. There is therefore a need to indicate the various parameters of such systems which may be varied and controlled. For such a purpose, a matrix of such parameters – Illumination source, Modulation (optical), Processing (e.g., camera setting), Sensor (detector) (IMPS)– is useful, an example of which is given in Figure 2.4.

Figure 2.4 shows a rectangle with each side representing a system property which can be varied – that is, optical source, camera detector, optical modulation and chromatic detection parameters. The optical source may be in the form of a visual display unit (VDU) whose chromatic output (e.g., purple, white, blue) can be software controlled to produce a range of different colours and which can be supplemented if needed by light-emitting diodes.

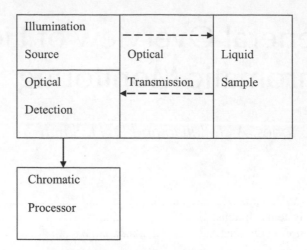

FIGURE 2.1   General structure of optical chromatic systems for liquid monitoring.

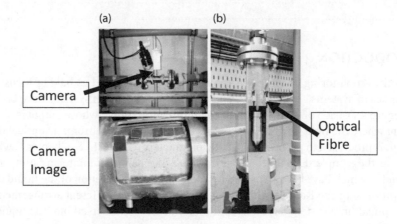

FIGURE 2.2   Examples of two options for chromatic monitoring of liquids on industrial sites: (a) camera-based system; (b) optical fibre-based system.

PARAMETER	PRINCIPLE	OPTICAL FIBRE	CAMERA
Voltage	Pockells Effrect	Y	?
Current	Faraday Effect	Y	?
Temperature	Thermochromic	Y	Y
Pressure	Photoelastic	Y	Y
Liquids / Tissue	**Multiple**	**Y**	**Y**
Position	Signature	?	Y
Vibration	Speckle Pattern	Y	?

FIGURE 2.3   Examples of condition monitored with chromatic optical fibres and electronic cameras.

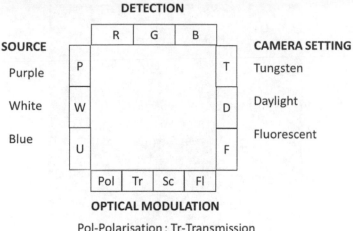

FIGURE 2.4 Permutations of adjustable operation parameters available for camera-based chromatic monitoring (IMPS).

The spectral sensitivity of the camera may be varied via a number of settings (e.g., tungsten, daylight, fluorescent) available on the camera.

The optical modulation can be arranged through the choice of VDU and camera settings plus the use of external filters (e.g., polarisation, fluorescence, etc.).

The chromatic output sensitivity can be adjusted by varying the amplitude of each camera output channel, that is, R, G, B.

These features illustrate the flexibility of such a system to be tuned for particular monitoring requirements.

2.3 ADDITIONAL MONITORING OPTIONS

A major aspect of the camera-based system is its production of a two-dimensional image which can be adapted for implementing additional features, as shown in Figure 2.5. This shows an image of skin tissue within a rectangular aperture window (WT) in a template around which there are a number of rectangular areas of different controlled colours. Two of these areas (r0, r2) represent chromatic signatures of the extreme conditions being monitored which can be used for normalising the chromaticity of the sample, WT. In addition, there are red-coloured dots at three of the template corners which are used for correcting the template orientation. Furthermore, there is a series of horizontal lines (top and bottom) for checking for the correct zoom level, along with a rectangular frame. These image features are all controllable in software via the camera.

2.4 FURTHER OPTIONS

A further option includes the system operation being controlled to selected extents via a personal computer (PC) connected to the system. If a small-volume unit is needed, a miniature VDU can be used for illumination and a Raspberry Pi unit used for operating and controlling the system with a capability of operating and controlling the system wirelessly from a remote location.

2.5 ILLUSTRATIVE EXAMPLES OF CHROMATIC SYSTEM EVOLUTION

The various aspects indicated have evolved through the production of systems required by particular applications and which reflect the evolution path to date. An initial system was capable of monitoring

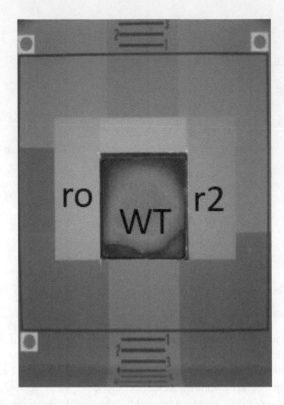

FIGURE 2.5 Typical camera image of a chromatically monitored sample (WT) including reference areas (r0, r2), spot orientation indicators and zoom correcting scale.

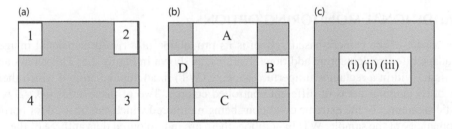

FIGURE 2.6 Choice of operational components for optical chromatic monitoring. (a) Images (1, 2, 3, 4) of a liquid in a cuvette produced with different VDU illumination (e.g., Chapters 4, 5, 7). (b) Choice of operating conditions (IMPS; Figure 2.4) [A = Processor response (R, G, B); B = Camera setting (e.g., tungsten, daylight etc.); C = Form of optical modulation (transmission, reflection, fluorescence, polarisation); D = Source colour (purple, white, blue etc.) (e.g., Chapter 7)]. (c) Three test samples [(i), (ii), (iii)] viewed through a central window in a peripheral. Reference template (e.g., Chapter 6).

petroleum fuel at fuel stations using optical fibre-based probing (Chapter 3). A PC-based system for monitoring *E. coli* in urine samples provides a first insight into a camera-based system (Chapter 4). Such a portable PC system was adapted for monitoring honey samples in remote geographical areas, including the use of fluorescence (Chapter 5). The approach was further adapted for the *in vivo* monitoring of jaundice in the tissue of newborn babies using a mobile phone system (Chapter 6). A small portable system for the remote monitoring of high-voltage transformer oils was produced using a Raspberry Pi–based monitoring unit (Chapter 7).

2.6 SUMMARY

It has been explained how new forms of liquid and tissue monitoring based upon various manifestations of chromatic monitoring have evolved. The coloured image on the book cover illustrates examples of various options for chromatically addressing a variety of applications. Various sections of the image are indicated in the greyscale Figure 2.6.

REFERENCE

Rallis, I., Deakin, A., Spencer, J.W., and Jones, G.R. (2005) Novel sensing techniques for industrial scale biodigesters. *Proc. of SPIE 17th Int. Conference on Optical Fibre Sensing.* 5855, 110.

3 Optical Chromaticity for Petroleum Discrimination

J. W. Spencer

CONTENTS

3.1 INTRODUCTION

There have been a number of attempts (Workman, 1996) to develop an optical system for distinguishing between different fuels purchased from fuel forecourts. Such a system would assist in preventing cross-contamination between diesel and petrol during delivery, ensure that a particular brand of fuel, at a higher price, is dispensed into customers' tanks and also stop contaminated fuel from being dispensed into a tank. The complexity of the fuel mix arising from the use of additives (~a few percent) such as cleaning agents makes the discrimination between different fuels difficult.

This provides a challenge for the application of liquid chromatic monitoring for distinguishing between different fuel types and brands and also for the fuel processing refinery. It would lead to the chromatic approach having the potential for producing sensors and sensor systems that could be used in fuel forecourts to monitor fuel quality in real time.

3.2 CHROMATIC MONITORING SYSTEM

A chromatic monitoring system for addressing samples of petroleum has been developed based upon polychromatic optical fibre transmission of light. Light from a white light emitting diode (LED) was transmitted through an optical fibre bundle to a remote probe which addressed a petroleum sample before returning the petroleum-modulated polychromatic light to a miniature optical spectrometer (Figure 3.1a). The spectrum produced by the spectrometer was addressed with three chromatic processors to yield outputs of R, G, B (Chapter 1) from which values of various chromatic parameters (x, y, z, L, H, S; Chapter 1) were produced. Figure 3.1b shows an image of the optical fibre probe, whilst Figure 3.1c shows the deployment of such probes on board a petroleum monitoring vehicle (courtesy of Fairbanks) for visiting various fuel filling stations for on-site fuel testing and on a fuel tank filler inlet (Figure 3.1d).

3.3 CHROMATIC ANALYSIS OF PETROLEUM SPECTRA

Figure 3.2 shows examples of the complex spectra of several diesel and petrol fuels overlaid on each other. These spectra indicate the high degree of similarity which exists between the spectra (300–1755 nm) of many different fuels. However, close inspection shows some possible differences

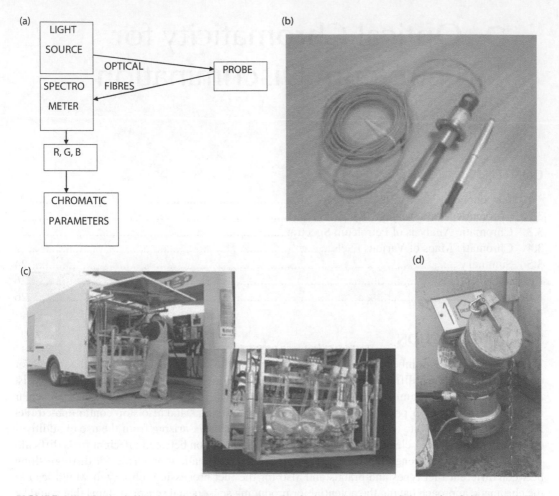

FIGURE 3.1 Optical fibre-based chromatic system for fuel monitoring. (a) Schematic of the optical fibre system. (b) Photograph of an optical fibre probe. (c) Photograph of a vehicle for monitoring at fuel-filling stations. (d) Optical fibre probe attached onto a fuel tank.

in the wavelength range 1138–1258 nm. Thus, chromatic processors (also shown in Figure 3.2) may be effectively deployed in this range to extract values of chromatic parameters for distinguishing the various fuel samples.

The outputs from the RGB filters were processed to yield values of the chromatic parameters H, S, L using the relationships given in Chapter 1.

3.4 CHROMATIC MAPS OF VARIOUS FUELS

Examples of H–S and H–L chromatic maps for 13 diesel and petrol samples (Fairbanks) are shown in Figure 3.3a and b. These results show that in both H–S and H–L maps, the diesel and petrol samples form two clear clusters, mainly governed by the H values.

Ultimate and premium samples of petrol obtained from retailers (some were well-known worldwide brands and others were local supermarket brands available only in the United Kingdom) have also been chromatically analysed. More than one spectrum was obtained for each sample type to reduce the noise. The fuels collected from these forecourts were processed at four different plants from the United Kingdom. The HS and HL values were processed from the R, G, B data for the same spectral range (1138 and 1258 nm). The HS values obtained show a separation of the two types

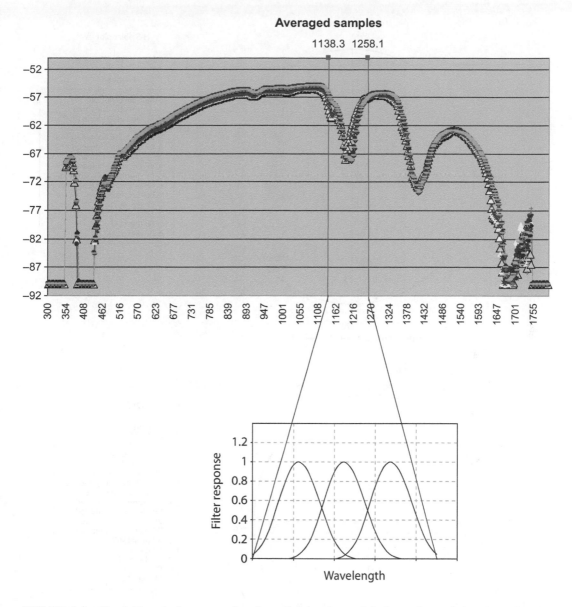

FIGURE 3.2 Overlaid optical spectra of various diesel and petrol fuel samples and the spectral range addressed by three chromatic processors.

of petrol (ultimate and premium) on an expanded chromatic map (Figure 3.4). Sample-to-sample variation was also observed.

Figure 3.4 also shows that a third cluster of samples is located between the two main clusters. These were premium fuels from one particular refinery (Lindsey).

Further chromatic data evaluation involved taking an average HS value for the various samples of each fuel type and presenting it as a single data point on an expanded H–S map (Figure 3.5). This reduced the number of data points in the figure to allow further observations.

It shows that for ultimate fuels, there is a good discrimination between the manufacturers of each product, although the spread for one manufacturer is more notable than others (i.e., Total). For premium fuels, there is a greater variation in the HS values associated with branded fuel.

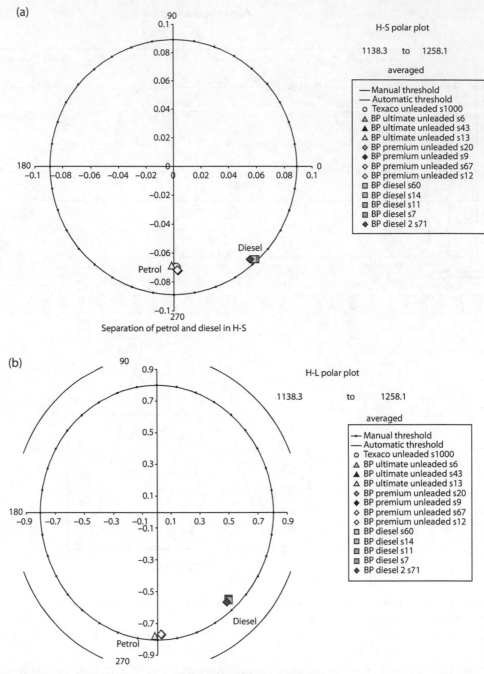

FIGURE 3.3   Chromatic H, S, L polar maps for diesel and petrol samples. (a) H–S map. (b) H–L map.

3.5   SUMMARY

The results of the investigation show the extent to which chromatic monitoring has the capability of distinguishing different fuel samples with similar complex optical spectra. It also demonstrates how optical chromaticity may be conveniently deployed using readily available cost-effective

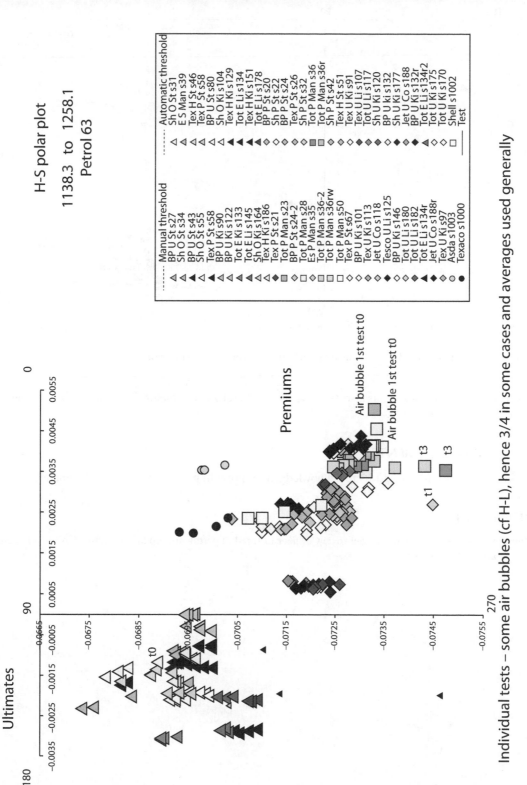

FIGURE 3.4 Expanded scale HS values for premium and ultimate fuel samples.

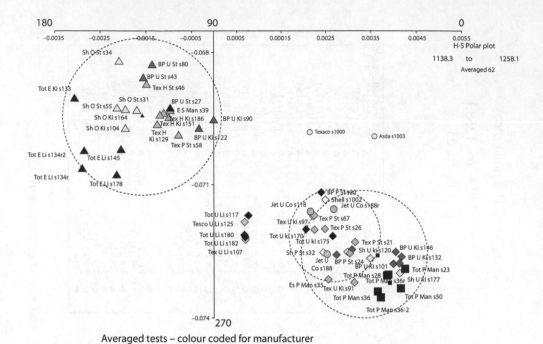

Averaged tests – colour coded for manufacturer

FIGURE 3.5   HS values for distinguishing fuel samples from different manufacturers.

instrument components. It has also been shown that diesel and petroleum fuels can be chromatically distinguished (Figure 3.3).

ACKNOWLEDGEMENTS

Fairbanks Environmental Ltd. are acknowledged for providing access to many fuel samples.

REFERENCE

Workman, J. (1996) Review: A brief review of near infrared in petroleum product analysis. *J. Near Infrared Spectrosc.* 4, 69–74.

4 PC-Based Chromatic Monitoring of *E. coli* in Urine

A. T. Sufian and G. R. Jones

CONTENTS

4.1 INTRODUCTION

Chromatic monitoring has been used for preliminary testing of infection in urine using a portable system. The system was based upon a combination of a webcam with a personal computer (Deakin et al., 2014) to provide a convenient and economic means for primary medical care. The approach has been shown to be capable of characterising complex samples of urine and its components.

4.2 CHROMATIC SYSTEM

The chromatic system consisted of optically addressing a urine sample in a transparent cuvette with polychromatic illumination produced by a tuneable laptop computer visual display unit (VDU) screen. The cuvette was placed between the VDU screen and a webcam which captured an image of the cuvette and computer screen (Figure 4.1a). The image captured by the webcam was transferred directly to the host laptop computer for recording and chromatic analysis. The VDU screen illumination could be tuned and controlled via the host computer. It involved adjusting the gains of the R, G, B screen channels to optimise performance via trial and error, which led to gains of 0.7 for R, G and B for this particular application. The webcam was adjusted so that the full illumination area provided on the VDU screen filled the camera image area. The screen areas on either side of the cuvette were tuned via trial and error for providing a means for any image illumination correction and to stabilise the webcam response.

Tests were performed with polychromatic light from the VDU screen transmitted through the cuvette and sample in the presence of ambient light and also with the VDU illumination replaced by a black card so that only ambient light was present. Figure 4.1b shows examples of images of the cuvette only captured by the webcam. Figure 4.1b (i) TRANS is the image of a normal urine sample with VDU transmission, whilst Figure 4.1b (i) REFL is the same sample with ambient light in the absence of VDU illumination. Figure 4.1b (ii) TRANS and REFL are the corresponding images with a highly contaminated urine sample. R, G, B outputs for the urine sample with and without VDU illumination were extracted from the images for further processing.

4.3 CHROMATIC MAPS AND PROCESSING

Figure 4.2 shows the optical transmission spectra of uncontaminated urine (Vekatratnam and Lents, 2011) and unspecified *E. coli* in sterilised de-ionised water (Alupoaei et al., 2004) over the wavelength

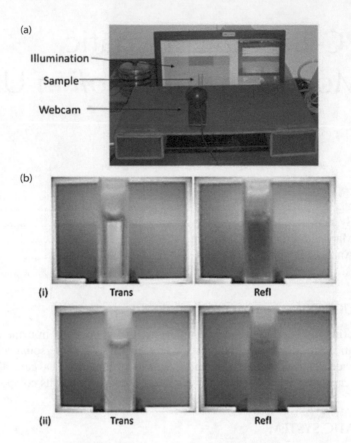

FIGURE 4.1 Portable computer-based chromatic monitoring system. (a) Layout of the PC monitoring system with cuvette illumination source, sample and webcam. (b) Examples of images of the monitored urine samples with illumination. (i) Negative UTI sample. (ii) Positive UTI sample. Trans – with transmitted screen Illumination; Refl – ambient light, no screen illumination (Deakin et al., 2014).

range 200–800 nm. Also shown in Figure 4.2 are the responses R, G, B of human vision receptors on which electronic camera detectors are based, covering only the wavelength range 400–650 nm. There remain differences between the two spectra in the 400–650 nm range.

The chromatic parameters used for monitoring urine samples by Deakin et al. (2014) are R and B, leading to chromatic maps of R: B. Since the system needed to have the capability for use under normal ambient illumination, it was necessary to obtain images first with VDU illumination, followed by images without VDU illumination [e.g., Figure 4.1b (i) and (ii)]. Thus, the sample transmission parameters were

$$R(TRANS) = R(VDU \text{ illumination}) - R(REFL) \tag{4.1}$$

$$B(TRANS) = B(VDU \text{ illumination}) - B(REFL) \tag{4.2}$$

The resultant VDU parameters were further normalised with respect to a water sample to provide a scale range (0–1), that is

$$R(TRANS)n = R(TRANS)/R(TRANS \text{ water}) \tag{4.3}$$

$$B(TRANS)n = B(TRANS)/B(TRANS \text{ water}) \tag{4.4}$$

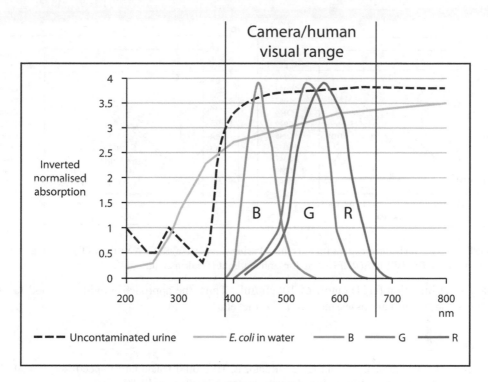

FIGURE 4.2 Optical spectra of uncontaminated urine and *E. coli* in water (Deakin et al., 2014).

whilst the ambient light parameters were likewise normalised

$$R(REFL)n = R(REFL)/R(REFL \text{ water}) \tag{4.5}$$

$$B(REFL)n = B(REFL)/B(REFL \text{ water}) \tag{4.6}$$

The chromatic maps used for checking the urine samples were R(TRANS)n versus B(TRANS)n and R(REFL)n versus B(REFL)n.

4.4 CHROMATIC ANALYSIS RESULTS

An illustration of test results for ten clinical urine samples is given in Figure 4.3a for VDU illumination (R(TRANS)n : B(TRANS)n) and Figure 4.3b for reflection (R(REFL)n : B(REFL)n).

The VDU illumination map shows that the clinical urine samples lie above the R(TRANS) n = B(TRANS)n locus, whereas the *E. coli* in the pure urine locus lies below the R(TRANS) n = B(TRANS)n locus. Comparison with known bacterial growth urine samples indicates that urine with bacterial growth </=10*5 cfu/mL lies in the NG sector of Figure 4.3a.

The ambient light (scattered light) map shows results for the ten urine samples tested along with the *E. coli* pure urine locus. These results show that all sampled points lie above the R(REFL) n = B(REFL)n locus. Significant bacterial growth samples lie in the G sector.

Chromatic results for 200 urine samples were reported and compared with urine culture results. The comparison showed that the number of true positive infection results was 73, and the number of true negatives was 66, that is, a total of 139 correct results. The number of false positives was 55 and false negatives 6, that is, only 6 samples from the 200 (3%) were incorrectly indicated as not being infected. Of the 55 false positive samples, 49 had some bacterial growth, albeit below the chosen

(a)
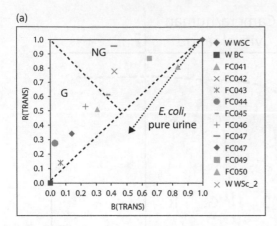

(b)
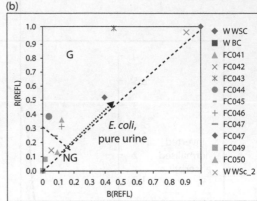

FIGURE 4.3 R : B chromatic maps for ten clinical urine samples. (a) R(TRANS) : B(TRANS) map for VDU illumination. (b) R(REFL) : B(REFL) map for ambient light (Deakin et al., 2014).

urinary tract infection (UTI) cutoff of 10^5 cfu/mL. Thus, the approach provided a good fail-safe capability for preliminary assessment of *E. coli* levels.

REFERENCES

Alupoaei, C. E., Olivares, J. A., and Garcia-Rubio, L. H. (2004) Quantitative spectroscopy analysis of prokaryotic cells, vegetative cells and spores, biosens. *Bioelectron.* 19, 893–903.

Deakin, A. G., Jones, G. R., Spencer, J. W., Bongard, E. J., HGal, M., Sufian, A. T., and Butler, C. C. (2014) A portable system for identifying urinary tract infection in primary care using a PC-based chromatic technique. *Physiol. Meas.* 35, 793–805.

Vekatratnam, A. and Lents, N. H. (2011) Zinc reduces the detection of cocaine, methamphetamine, and THC by ELISA. *Urine Testing J. Anal. Toxicol.* 35, 333–340.

5 Chromatic Monitoring of Honey Samples

A. T. Sufian and G. R. Jones

CONTENTS

5.1 INTRODUCTION

Honey is a natural sweet substance produced by honeybees worldwide (Figure 5.1). It consists of a complex mixture of water, various sugars, amino acids, enzymes, vitamins and minerals (White, 1975). The composition is influenced not only by natural factors (geographical origins, botanical sources, environment, climate) (Anklam, 1998) but also by processing, handling and storage (Martin and Bogdanov, 2002). Monitoring such a complex liquid is important to the food industry (Bogdanov et al., 1999; Martin and Bogdanov, 2002; Pilizota and Tiban, 2009) to ensure genuine quality and to identify fraudulent imitations and adulteration which affect the integrity of the honey market (Pilizota and Tiban, 2009; El-bialee et al., 2013; Roshan et al., 2013).

Low-cost, traditional methods (taste, smell, visual observation) are not sufficiently reliable for identifying fraudulent and adulterated honeys, whilst sophisticated optical measurement techniques (Gonzales et al., 1999; Isengrad et al., 2001; Terrab et al., 2002; Nanda et al., 2003; El-bialee et al., 2013; Özbalci et al., 2013; Roshan et al., 2013) require expensive instrumentation (e.g., spectroscopic, colorimetric, polarimetry, refractometry) and are not easily adapted for field use. Methods based upon physicochemical procedures are time consuming and require a multiplicity of tests with expensive equipment and laboratory infrastructure (Anklam, 1998; Bogdanov et al., 1999; Martin and Bogdanov, 2002).

However, optical chromaticity provides a cost-effective means for the preliminary optical monitoring of honey samples with a portable computer (PC) being deployable at remote honey-producing areas as well as consumer sources (Sufian, 2014). Chromatic methods have been deployed for the combined analysis of optical transmission, polarisation and fluorescence signals from various honeys and for producing chromatic maps from which the honey condition may be checked. Such a PC-based system is described, along with chromatic processing to provide a preliminary indication of honey quality.

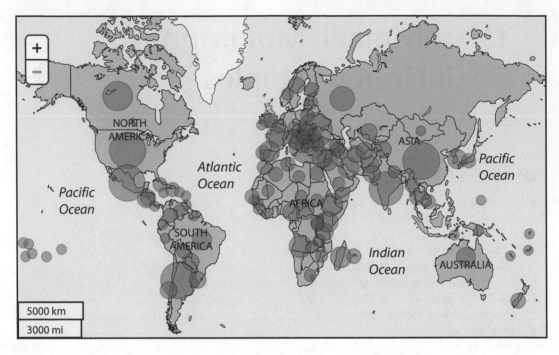

FIGURE 5.1   Map of distribution of world production of honey as of FAO Statistics 2018.

5.2   PORTABLE COMPUTER-BASED CHROMATIC MONITORING SYSTEM

The chromatic monitoring system for monitoring honey samples is a more sophisticated form of the PC-based urine monitoring system (Chapter 4). It involves not only chromatic optical transmission but also chromatic fluorescence and polarisation. Figure 5.2a shows a photograph of such a system, whilst Figure 5.2b shows a schematic diagram of the system layout. A honey sample is contained in a transparent cuvette located between a visual display unit (VDU) screen of a portable computer (which is one illumination source) and a webcam (which detects the honey image). The VDU screen, webcam and computer were interconnected to form the basis of the compact, self-contained, portable and cost-effective unit.

Optical transmission tests were performed with polychromatic light produced by the VDU screen and controlled via the computer software (Chapter 4). An image of the cuvette and surrounding screen area was captured by the webcam (Figure 5.3a) and transferred to the computer for chromatic processing.

Optical polarisation tests were performed with a polarisation filter positioned between the VDU screen and the honey-containing cuvette, whilst a second polarising filter with its polarising axis suitably rotated was placed between the cuvette and webcam (Figure 5.2).

Optical fluorescence was observed with two light-emitting diodes (LEDs) to provide short-wavelength monochromatic light (405 nm wavelength) for addressing the honey-containing cuvette (Figure 5.2). The two LEDs were inclined at 45 degrees to the camera line of sight on the camera side of the cuvette so that back-scattered fluorescent light could be captured by the webcam together with the VDU illumination just at either side of the sample, while VDU light transmitted through the sample was blocked using a black card.

5.3   TEST RESULTS

Examples of images captured with the system for a honey sample are shown in Figure 5.3. An image obtained with polychromatic white from the VDU screen to provide transmission information is shown in Figure 5.3a, whilst an image with polarised white light from the VDU is shown in Figure

(a)   (b)

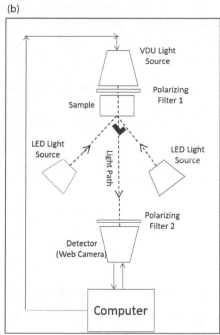

FIGURE 5.2 PC/webcam-based chromatic system for monitoring honey samples. (a) View of overall system. (b) Schematic diagram (Sufian, 2014).

5.3b and an image with fluorescence produced by the honey sample with the light from the LEDs is shown in Figure 5.3c.

R, G, B values were obtained for chosen test X(w) = R(w), G(w) or B(w) and reference X(ref) = R(ref), G(ref) or B(ref) areas in such images (e.g., Figure 5.3a) for transmission, polarisation and fluorescence illumination. A correction factor X(cf) was applied to accommodate test-to-test variations, yielding test result X(w)s.

$$\text{Correction Factor } X(cf) = R(cf), G(cf) \text{ or } B(cf) = [X(ref)set]/[X(ref)]$$

$$\text{Corrected test result } X(w)s = R(w)s, G(w)s \text{ or } B(w)s = [X(w) \times X(cf)]$$

where X(ref)set is the chosen set values of X(ref).

(a)   (b)   (c)

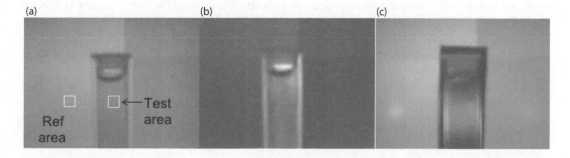

FIGURE 5.3 Examples of images of a honey sample and background VDU screen under different optical conditions. (a) Optical transmission. (b) Polarised light. (c) Optical fluorescence (Sufian, 2014).

In addition, changing ambient conditions were accommodated via X(A) by placing a black card behind an empty cuvette and extracting RGB values in a similar manner to the test samples. The test results X(w)s were normalised according to samples corresponding to the maximum X(w)smax and minimum X(w)smin levels of the measurement ranges.

$$\text{Normalised test result } X(w)N = [X(w)s - X(A)]/[X(w)smax - X(w)smin]$$

The previous procedures were applied to each of the optical signals, that is, transmission, polarisation and fluorescence.

For transmission, X(w)max and X(w)min corresponded respectively to water and a highly turbid honey sample being screen-illuminated in the cuvette.

For polarised light, X(w)max and X(w)min corresponded respectively to a highly concentrated sugar solution and water in the cuvette being addressed by the screen illumination via two polarising filters (Figure 5.2) aligned at 45° to each other.

For fluorescence, X(w)max and X(w)min corresponded to a high-purity honey sample and water in the cuvette being addressed by the LED light.

5.4 CHROMATIC INTERPRETATION OF TEST RESULTS

The normalised test data (transmission, polarisation, fluorescence) for 21 different honey samples from various sources and various degrees of purity, quality and so on were processed to display chromatically for qualitative visual comparison as well as on chromatic maps.

Chromatic cluster maps of the processed honey results were formed as graphs of R:B and G:B. Such maps were produced for the transmission (Rt:Bt), polarisation (Rp:Bp) and fluorescence (Gf:Bf) data (Figure 5.4a–c). Based upon calibration results with mixtures of honey, water and syrup, sector boundaries were produced on these maps to classify honey samples from different sources and conditions. The sectors on the transmission map (T1–T5) distinguished clear from turbid honeys; on the polarisation map (P–P3) low, high-sugar and sugar-diluted honeys, whilst the Fluorescence map (F1–F2) distinguished pure (F1) and impure (F2) honeys. Clustering of various honey samples on the three chromatic maps (transmission, polarisation, fluorescence) was interpreted as follows.

5.4.1 TRANSMISSION R:B MAP

Region T1 honey samples were optically transparent due to a high liquid content, indicating water dilution during processing and exposure to overheating or overfiltering, that is, processed/adulterated honeys. See Figure 5.4a.

Region T2, T3 honey samples were moderately clear (T2) or had low turbidity (T3), and as such were raw honeys with low or moderate pollen contents, depending upon the floral source (Al-Zoreky et al., 2001).

Region T4 honey samples were highly turbid with lower light transmission, which suggested a long shelf life and high storage temperature (White, 1975; Gonzales et al., 1999), leading to chemical instability, botanical source or crystallisation.

Region T5 honey samples showed little optical transmission or reflection, possibly due to their mineral content (González-Miret et al., 2005).

5.4.2 POLARISATION R:B MAP

Region P1 honey samples had relatively high levels of polarised short- and long-wavelength light caused by high sugar levels (glucose/fructose) with water content. See (Figure 5.4b).

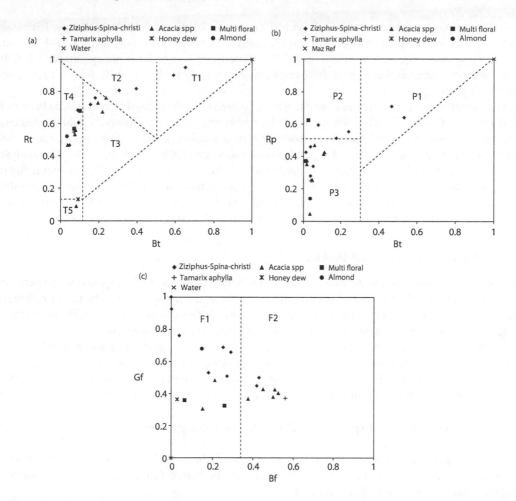

FIGURE 5.4 Primary chromatic maps for various honey samples (------- classification boundaries) (a) Transmitted light parameter Rt:Bt. (b) Polarised light parameter Rp:Bp. (c) Fluorescence parameter Gf:Bf (Sufian, 2014).

Region P2 honey samples had a higher fraction of polarised long-wavelength light due to glucose/fructose being major constituents (Al-Zoreky et al., 2001).

Region P3 honey samples had relatively low levels of polarised light, indicating low sugar (glucose/fructose) content.

5.4.3 FLUORESCENCE G:B MAP

Region F1 honey samples (which were of moderate purity) had fluorescence spectra shifted towards the medium wavelengths (G) due to higher ash content (Al-Zoreky et al., 2001) and are traditionally regarded as genuine/good-quality honey (i.e., Grade 1). See (Figure 5.4c).

Region F2 honey samples had high levels of impurities and high water content. F2 samples, which also lie in Regions T4 and T5 of Figure 5.4a, had been overheated or had invert sugar (cane sugar).

5.5 QUANTIFICATION OF PERFORMANCE

The chromatic signatures of a honey sample on each of the three chromatic maps (Figure 5.4a–c) were combined to provide an overall empirical quantification of the honey quality by allocating a quality score to each of the different map sectors. Each sector, T1–T5, on the transmission map (Figure 5.4a) carried a score 0–1, with 0 for poor quality and 1 for pure honey. Likewise, sectors P1–P3 on the polarisation map (Figure 5.4b) were scored 0–1, and sectors F1–F2 on the fluorescence map (Figure 5.4c) were scored 0–2. The three scores for a honey sample (T, P, F) from each map were added to give an overall quality indication (T + P + F), with 0 being poor and 4 being very good.

The overall score for each honey sample was compared with the quoted formal grade (Poor = 1, Moderate = 2, Good = 3) of the honey provided by the honey producer. The honey quality predicted chromatically was compared statistically with the formal grading in terms of sensitivity, specificity, positive predictive value (PPV) and negative predictive value (NPV) (Appendix 5A). The analysis showed that high-quality honey samples could be identified with a sensitivity of 91%, a specificity of 80%, a positive predictive value of 83% and a negative predictive value of 89%, and that poor-quality honey samples were identified with a sensitivity of 75%, a specificity of 92%, a positive predictive value of 86% and a negative predictive value of 86% (Sufian, 2014).

5.6 OVERVIEW AND SUMMARY

Chromatic techniques have been used for monitoring honey quality and adulteration using a personal computer, its visual display unit as one illumination source and a connected webcam for capturing the chromatic signatures of a honey sample. The system is flexible in being operable in the three optical domains of light, transmission/absorption, polarisation and fluorescence.

Chromatic maps for each of these three domains were produced and calibrated with mixtures of water, syrup and honey before being used for discriminating between various honey samples.

Calibration, normalisation and ambient light compensation procedures were developed to allow operation under a range of illumination conditions such as in the field (sunlight) and at commercial premises.

The approach was shown to perform well via a statistical comparison of results with formal honey classification.

The self-contained, flexible, cost-effective nature of the system and its portability enabled preliminary tests to be undertaken in remote rural areas of Yemen (Sufian, 2014), where several of the honey samples were tested (Figure 5.5).

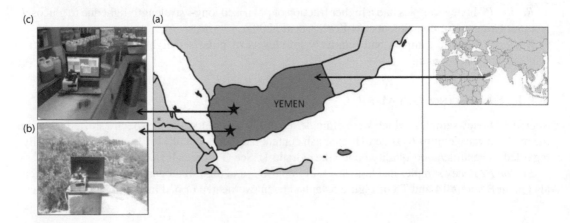

FIGURE 5.5 PC-based honey monitoring system used onsite in Yemen. (a) Geographical map of Yemen with the field-test sites; (b) onsite honey monitoring with a PC in Tubasha'a Village, Sabir Mount, Taiz governorate – Midlands, Yemen; (c) onsite honey monitoring with a portable PC in a local honey shop, Sanaa, North – Yemen (Sufian, 2014).

ACKNOWLEDGEMENTS

The supply of honey samples by Dr. Eida Alssadi from the Center of Food and Medicine in Sana'a, Yemen, is acknowledged. Support provided by the British-Yemeni Society for field tests is also appreciated.

APPENDIX 5A: FORMULAE FOR SENSITIVITY, SPECIFICITY, POSITIVE PREDICTED VALUE, NEGATIVE PREDICTED VALUE

Sensitivity (Sn), specificity (Sp), positive predicted value (PPV) and negative predicted value (NPV) are defined by the following equations (e.g., Deakin et al., 2014).

$$\text{Sensitivity (Sn)} = TP/(TP + FN)$$

$$\text{Specificity (Sp)} = TN/(TN + FP)$$

$$\text{Positive Predicted Value (PPV)} = TP/(TP + FP)$$

$$\text{Negative Predicted Value (NPV)} = TN/(TN + FN)$$

where TP and FP are respectively the number of true and false positive chromatic results, and TN and FN are respectively the number of true and false negative chromatic results.

REFERENCES

Al-Zoreky, N., Alza'aemy, A. and Alhumiari, A. 2001. Quality spectrum of Yemeni honey. *Journal of Agricultural Science, Damascus University*, 17(2), p. 110.

Anklam, E. 1998. A review of the analytical methods to determine the geographical and botanical origin of honey. *Food Chem*, 63(4), pp. 549–562.

Bogdanov, S. et al. 1999. Honey quality and international regulatory standards: Review by the international honey commission. *Bee World*, 80(2), pp. 61–69.

Deakin, A.G., Jones, G.R., Spencer, J.W., Bongard, E.J., Gal, M., Sufian, A.T. and Butler, C.C. 2014. A portable system for identifying urinary tract infection in primary care using a PC-based chromatic technique *Physiol. Meas*. 35(5), pp. 793–805.

El-bialee, N., Rania, K.I., El-bialee, A., Harith, M.A. and Darwish, N.E.M. 2013. Discrimination of honey adulteration using laser technique. *Australian Journal of Basic and Applied Sciences*, 7(11), pp. 132–138.

Gonzales, A.P., Burin, L. and Buera, M.D.P. 1999. Color changes during storage of honeys in relation to their composition and initial color. *Food Res Int*, 32(3), pp. 185–191.

González-miret, M.I., Heredia, F.J., Terrab, A., Hernanz, D. and Fernández-recamales, M.A. 2005. Multivariate correlation between color and mineral composition of honeys and by their botanical origin. *J Agric Food Chem*, 53(7), pp. 2574–2580.

Isengrad, H., Schultheiss, D., Radović, B. and Anklam, E. 2001. Alternatives to official analytical methods used for the water determination in honey. *Food Control*, 12(7), p. 459.

Martin, P. and Bogdanov, S. 2002. *Honey Authenticity: A Review*. Swiss Bee Research Centre, Dairy Research Station, Liebefeld; Q. P. Services, Hayes, Great Britain.

Nanda, V., Sarkar, B.C., Sharma, H.K. and Bawa, A.S. 2003. Physico-chemical properties and estimation of mineral content in honey produced from different plants in Northern India. *J Food Compos Anal*, 16(5), pp. 613–619.

Özbalci, B., Boyaci, I.H., Topcu, A., Kadilar, C. and Tamer, U. 2013. Rapid analysis of sugars in honey by processing Raman spectrum using chemometric methods and artificial neural networks. *Food Chem*, 136(3–4), pp. 1444–1452. [19]

Pilizota, V. and Tiban, N. 2009. Advances in Honey Adulteration Detection. [Food Safety Magazine], [online]. Available: http://www.foodsafetymagazine.com/magazine-archive1/augustseptember-2009/advances-in-honey-adulteration-detection/

Roshan, A. et al. 2013. Authentication of monofloral Yemeni sidr honey using ultraviolet spectroscopy and chemometric analysis. *J Agric Food Chem*, 61(32), pp. 7722–7729.

Sufian, A.T. 2014. Honey monitoring in the Yemen. *British-Yemeni Society Journal*, 22, pp. 50–57.

Terrab, A., Diez, M.J. and Heredia, F.J. 2002. Chromatic characterisation of Moroccan honeys by diffuse reflectance and tristimulus colorimetery- non-uniform and uniform colour spaces. *Journal of Food Science and Technology International*, 8(4), pp. 189–195.

White, J.W. 1975. *Composition of Honey. Honey: A Comprehensive Survey.* Heinemann: London, UK, pp. 157–206.

6 Mobile Phone Chromatic Monitoring of Jaundice *In Vivo*

A. T. Sufian, G. R. Jones and H. M. Shabeer

CONTENTS

6.1 INTRODUCTION

The occurrence of jaundice (bilirubin) in newborn babies can lead to serious mental deficiencies in the baby. The established procedure for diagnosing bilirubin is by extracting and analysing a blood sample from a baby (invasive blood sampling). A possible alternative to such invasive blood sampling is the use of transcutaneous bilirubinometry, which involves optically monitoring the skin tissue to avoid the need for disruptive blood extraction. The first transcutaneous bilirubinometer was introduced in 1980 (Yamamanouchi et al., 1980). Since then, several other devices have been developed, and important adjustments (such as the correction for the presence of other skin chromophores [e.g., melanin, haemoglobin]) have been made to improve their accuracy. These second-generation bilirubinometers are suitable for the screening of hyperbilirubinaemia, leading to a considerable decrease of the number of hospital re-admissions. However, after more than 30 years of development, no transcutaneous bilirubinometer has proven itself a worthy replacement for invasive blood sampling. Reasons for this limited clinical value are diverse (e.g., the technological design of the bilirubinometers, the method of clinical evaluation and variations between patients) but have not been investigated thoroughly in the literature.

A chromatically based approach has therefore been investigated in a new form of bilirubinometer. The approach is based upon a light-tight, handheld unit incorporating a mobile phone, an optical referencing template and sample window illuminated with white-light light-emitting diodes (LEDs) (Sufian et al., 2018). Various optical chromatic corrections are conveniently incorporated, making the unit suitable for use in remote locations which have mobile phone system access.

6.2 CHROMATIC MONITORING UNIT

The monitoring unit is based upon chromatically addressing white light from a group of light-emitting diodes reflected/scattered from the skin tissue of a neonate *in vivo* plus optical references using a mobile phone camera. Figure 6.1 shows such a monitoring unit. Figure 6.2 shows a schematic diagram of the overall system.

The system consists of a removable mobile phone camera mounted upon a light-tight housing for monitoring a reference template carrying a rigid plastic window at the other end of the light-tight enclosure. The rigid plastic window needs to be pressed against the skin tissue under test (e.g.,

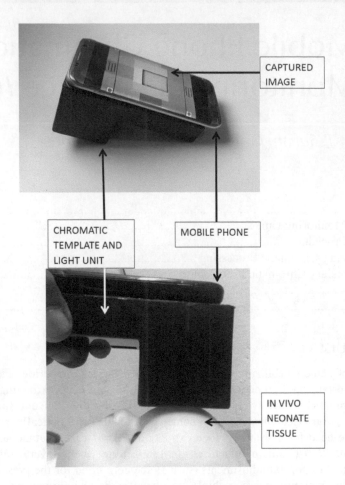

FIGURE 6.1 Light-tight handheld unit with mobile phone.

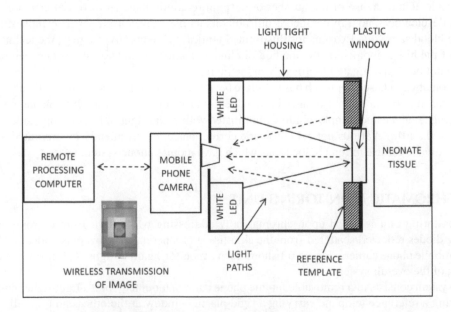

FIGURE 6.2 Schematic diagram of the handheld chromatic unit.

FIGURE 6.3   Template images captured by a mobile phone camera [W0 (set Bl = 0), W2(set Bl = 10 mg/dL) and WT (neonate nose tissue) in the sample window set reference ro = (set Bl ≈ 0), r2 = (set Bl ≈ 10)].

the nose) in order to occlude blood vessels from the tissue and so enhance the optical influence of the tissue. The template, including the plastic window, is illuminated by an array of white-light surface-mounted diodes arranged in a strip (0.2 W) and circled around an aperture (Figure 6.2), through which the camera lens observes and captures an image of the template and window (Sufian et al., 2018).

The template consists of a colour-printed paper with a rectangular and rigid sample viewing plastic window (Figure 6.2). The printed paper has several sectors with different chromatic signatures (Figure 6.3), which are empirically chosen, along with some alignment-assisting features. Two template sectors adjacent to the plastic window have chromatic signatures chosen empirically from occluded skin images of two neonates, one without jaundice (ro) and one with 10 mg/dL bilirubin (r2). These areas provide an indication relative to a sample in the window as to whether the camera settings and so on vary so that compensation can be made. The remaining template areas are empirically coded to accommodate automatic image adjustments made by the camera operating system and can vary for different cameras. If the camera is operated in "auto mode", the exposure time is automatically set. Geometric alignment features on the template (Figure 6.3) (red dots for correct angular orientation, square black frame for image size) ensure that the camera automatically addresses the relevant template sectors (ro, r2, W) for chromatic analysis.

6.3   CHROMATIC PROCEDURES

6.3.1   Primary Chromatic Processing

Relevant areas of sectors on the image captured by the mobile phone camera (sample window [WT], reference sectors [ro, r2]; Figure 6.3) produce outputs for the three chromatic parameters R, G, B. The wavelength responses of R, G, B are non-orthogonal, as shown in Figure 6.4. They cover the wavelength ranges of various skin tissue components (bilirubin, haemoglobin, melanin; Bhutani et al., 2000), as indicated in Figure 6.4. Thus, melanin (Me) and haemoglobin (He) are covered by all three processors (R, G, B), whilst bilirubin is covered mainly by processor B and to some extent by G.

The R, G, B outputs may be processed to yield chromatic parameters X, Y, Z and L (Chapter 1), with X, Y, Z being represented on a two-dimensional chromatic map (Chapter 1). Some preliminary chromatic signatures of 48 different *in vivo* tissue samples with different levels of bilirubin obtained with several different mobile phone cameras are shown on the Y:Z map of Figure 6.5. This illustrates the complex distribution of data produced by such basic chromatic processing.

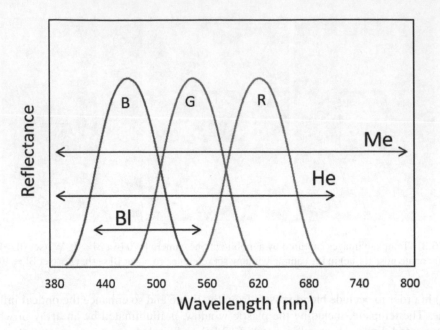

FIGURE 6.4 Typical overlapping wavelength responses (R, G, B) of an electronic camera relative to wavelength ranges of some skin tissue components (bilirubin [Bl], haemoglobin [He], melanin [Me]) (Sufian et al., 2018).

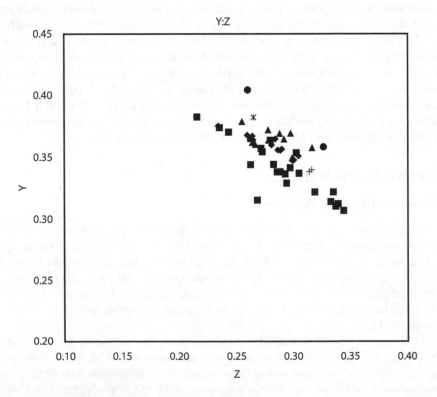

FIGURE 6.5 Basic Y:Z chromatic map of bilirubin test results

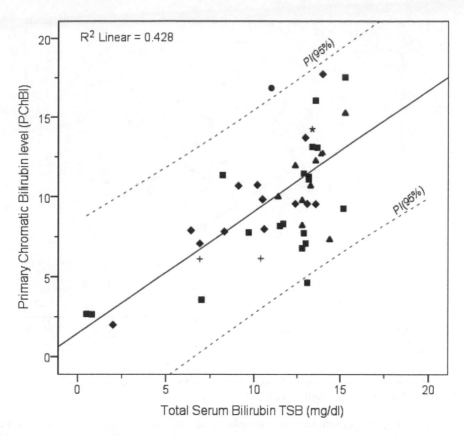

FIGURE 6.6 Primary chromatic bilirubin PCh(BL) versus total serum bilirubin [TSB (mg/dL)] (48 neonate test results, six different phone cameras; solid line → regression line, dashed lines → 95% prediction interval boundaries) (Sufian et al., 2018).

A basic chromatic calibration curve for bilirubin levels may be derived from the chromatic map of Figure 6.5 using a calibration parameter (Y/Z), which is a form of di-stimulus dominant wavelength (Sufian et al., 2018) obtained from the gradient of the locus from the origin to the Y, Z point.

A linear variation of this ratio between the set values for Bl = 0 and 10 mg/dL represented by the Y/Z values of reference papers Wo (Bl = 0) and W2 (Bl = 10 mg/dL) in the sample window may be assumed to provide an estimate of the bilirubin level [PCh(Bl)] from the measured value of a tissue (T) in the sample window [(Y/Z)(WT)], that is

$$PCh(Bl) = [(Y/Z)(WT) - (Y/Z)(Wo)]/[(Y/Z)(W2) - (Y/Z)(Wo)]$$

Bilirubin levels [PCh(Bl)] obtained with this equation for the 48 *in vivo* tissue samples with various cameras are shown in Figure 6.6 as a function of Bl values obtained for each neonate from blood tests yielding total serum bilirubin (TSB) levels.

The results have a high degree of scatter, which illustrates the complicating effects of various factors such as the use of different types of cameras, different tissue melanin levels in individual neonates and so on. Consequently, various correction procedures need to be incorporated through the use of secondary chromatic processing.

6.3.2 SECONDARY CHROMATIC PROCESSING

Secondary chromatic processing allows corrections to be made for differences between various cameras (e.g., different pixel densities [0.8–8.3 megapixels; Sufian et al., 2018]; different relative

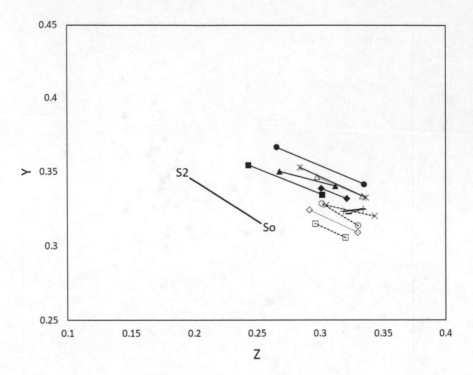

FIGURE 6.7 Chromatic Y:Z map showing locus of set reference areas (So–S2) and loci of samples in window (Wo–W2) for 12 different cameras (Sufian et al., 2018).

numbers of R, G, B pixels per unit area etc.), template and window distortions and so on. Such effects are compensated via the use of the reference areas on the template (ro, r2) (Figure 6.3) and window references (So, S2) via a Y:Z chromatic compensating map; see Figure 6.7. This shows a locus of the set values So–S2, along with the corresponding loci for each of 12 different cameras.

Such a map is used for producing correction factors (CFs) to account for various complexities by quantifying deviations from the set S0–S2 locus. A correction factor for Z is the ratio of a reference sample in the window (e.g., ZSo) to that of the set reference value (ZWo), that is

$$CF(Z, Wo) = (ZSo)/(ZWo)$$

A corrected value of a tissue sample [Z(WTo)] is then given by

$$Z(WTo) = Z(WT) \cdot CF(Z, Wo)$$

A similar expression applies with regard to the S2 reference, the choice of reference preferred depending upon the reference nearest to the measured sample value. Furthermore, a similar correction procedure is applied to the chromatic Y parameter.

A calibration graph of Z(WTo) versus the total serum bilirubin using the chromatic test results obtained with the 48 different tissue samples *in vivo* is shown in Figure 6.8a. Thus, the Z TSB value indicated by the corrected bilirubin level for a tissue sample with Z(WT)* is given by the TSB value ZCh(Bl) indicated by the calibration curve (solid line) (Figure 6.8a).

A corresponding calibration graph of Y(WTo) versus TSB is shown in Figure 6.8b, which likewise yields a corrected bilirubin level of YCh(Bl). The best estimate for a bilirubin level correction is given in practice by

$$SChBl = [YCh(Bl) + ZCh(Bl)]/2$$

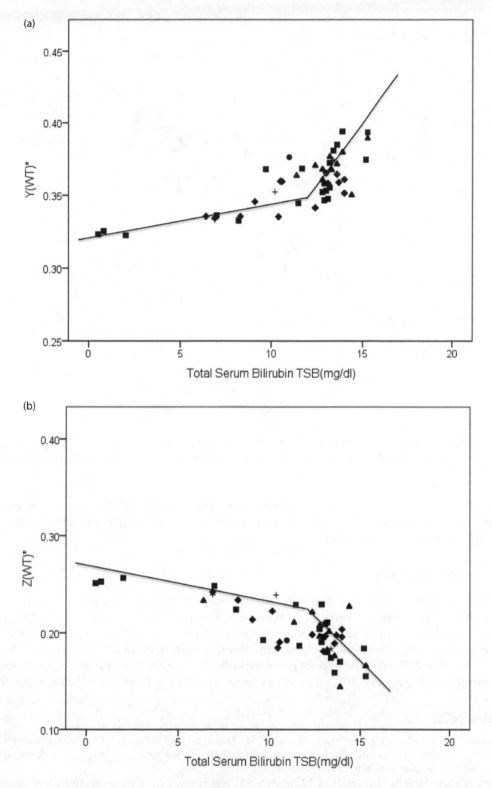

FIGURE 6.8 Calibration graphs for corrected (WT)*: as a function of total blood bilirubin level TSB (mg/dL). (a) Y(WT)*: TSB, (b) Z(WT)*: TSB (Sufian et al., 2018).

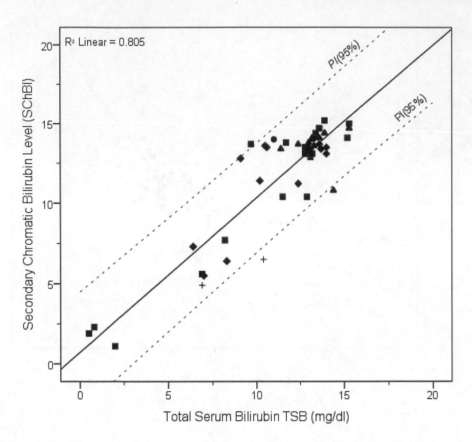

FIGURE 6.9 Corrected chromatic bilirubin [SCh(Bl)] versus total serum bilirubin (mg/dL) (48 neonate tests, six different phone cameras; solid line → regression line, dashed lines → 95% prediction interval boundaries) (Sufian et al., 2018).

Figure 6.9 shows values of the corrected chromatic bilirubin levels for the 48 *in vivo* tissue samples using the Y(WT)* and Z(WT)* calibration graphs of Figure 6.8. SCh(Bl) as a function of the TSB (mg/dL) was obtained from blood tests.

6.4   CONCLUSIONS

Compared with total serum bilirubin values (blood tests), the method of correcting the chromatic results has a significantly improved linear regression correlation coefficient (R^2) of 0.81 compared with a 0.43 value for the uncorrected chromatic results (Sufian et al., 2018). The chromatic method also compares favourably with test results reported by other investigators, for which $R^2 \approx 0.31$–0.71 (Sufian et al., 2018). The 95% prediction interval boundaries [PI(95%)] were improved from ±9 mg/dL for the uncorrected results (Figure 6.3) to ±3 mg/dL for the corrected results (Figure 6.9) (Sufian et al., 2018).

REFERENCES

Bhutani, V.K. Gourley, G.R., Adler, S., Kreamer, B., Dalin, C., and Johnson, L.H. (2000) Noninvasive measurement of total bilirubin in a multiracial predischarge newborn population to assess the risk of severe hyperbilirubinemia. *Pediatrics*, 106, e17.

Sufian, A.T., Jones, G.R., Shabeer, H.M. Elzagzoug, E.Y., and Spencer, J.W. (2018) Chromatic techniques for in vivo monitoring jaundice in neonate tissues. *Physioplogical Measurement*, 39(9), 095004.

Yamamanouchi, I., Yamanouchi, Y., and Igarashi, I. (1980) Transcutaneous bilirubinometry: Preliminary studies of noninvasive transcutaneous bilirubin meter in the Okayama National Hospital. *Pediatrics*, 65, 195–202.

7 Optical Chromatic Monitoring of High-Voltage Transformer Insulating Oils

A. T. Sufian, E. Elzagzoug and D. H. Smith

CONTENTS

7.1 INTRODUCTION

High-voltage transformers are monitored in order to avoid unnecessary electric power interruptions and to extend the transformer's service life (Lorin, 2005). One approach has been to monitor the electrically insulating oil in which the transformer is immersed using several techniques (Bureau of Reclamation et al., 2005), including optical ones such as colour index (CI) (Nadkarni, 2007). The use of chromatic techniques for monitoring changes in the optical spectra of polychromatic light transmitted through the oil has been shown to provide a versatile means to yield additional information to that obtained with CI and other techniques. A number of convenient-to-use, cost-effective systems for such chromatic monitoring have been investigated (Elzagzoug et al., 2014) which use different combinations of polychromatic light sources (white LED, visual display unit [VDU] screen) and optical detection units (spectrometer, mobile phone cameras). The use of a combination of a VDU screen and a mobile phone camera has been shown to be attractive in providing a tunable polychromatic light source with a user-versatile detection unit.

The optical techniques used are based upon the transmission of polychromatic light through the oils (Elzagzoug et al., 2014), observation of the fluorescence produced by monochromatic laser light (Lo et al., 2017) and a combination of transmission and fluorescence techniques (Sufian and Jones, 2017).

7.2 CHROMATIC OPTICAL TRANSMISSION

7.2.1 SYSTEM STRUCTURE AND OUTPUTS

The form of an optical transmission system for addressing an oil sample in a transparent cuvette is shown in Figure 7.1a. This indicates the interconnection between a polychromatic light source

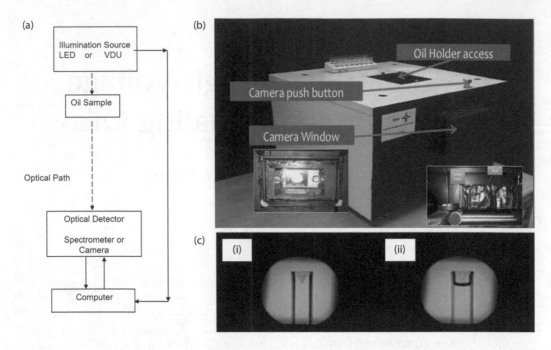

FIGURE 7.1 Optical transmission through transformer oils (Elzagzoug, 2013). (a) Schematic diagram of optical transmission system. (b) View of the portable chromatic oil monitoring system (PCOMS). (c) Typical cuvette images (i) with and (ii) without oil.

(e.g., LED, VDU), oil-containing cuvette, wideband optical detector (e.g., spectrometer, mobile phone camera) and data processing computer. A camera-based benchtop form of this portable chromatic oil monitoring system (PCOMS) (Figure 7.1b) has been used for laboratory development tests. A typical image obtained with the arrangement and a cuvette containing oil is shown in Figure 7.1c.

The possible permutations of source (LED, VDU) and detection units (spectrometer, mobile phone) which have been assessed (Elzagzoug, 2013) are shown in Table 7.1.

This has enabled the operation of the VDU–mobile phone camera system to be compared with a conventional measurement system (LED–spectrometer), consistent with the considerations of Chapter 2. The options for the camera operation and so on were based upon the illumination/modulation/processing/sensor (IMPS) diagram (Chapter 2); see Figure 7.2.

Although the transmission spectra obtained with a VDU source are not continuous, as are those obtained with a LED, the variation with different oil samples is similar. See Figure 7.3.

TABLE 7.1

Permutations of Sources and Detectors Tested

Source	Detector
LED	Spectrometer
LED	Mobile Phone Camera
VDU Screen	Spectrometer
VDU Screen	Mobile Phone Camera

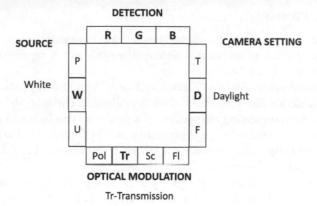

FIGURE 7.2 IMPS permutation of chromatic components properties for oil transmission tests (chosen operational parameters highlighted).

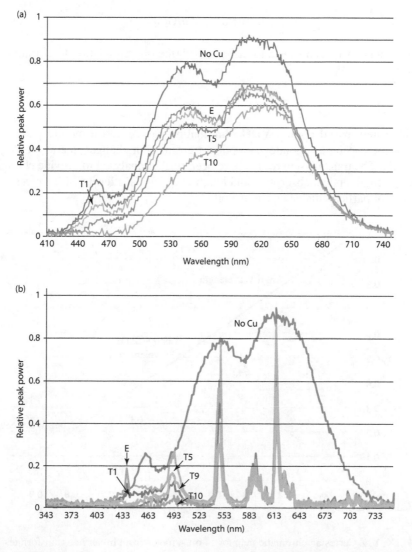

FIGURE 7.3 Transmission spectra of cuvette and various oil samples (a) LED source; (b) VDU.

7.2.2 Chromatic Processing

Chromatic processing of the R, G, B outputs yields XYZ chromatic maps (Chapter 1), with Z representing relative changes in the short-wavelength regions and X representing changes in the long-wavelength regions (Figure 7.4).

Tests with real transformer oils from Electricity North West (ENW) indicated that chromatic changes were detectable for different levels of oil degradation. The map shows that a convenient chromatic parameter for representing changes in oil degradation can be based upon the gradient of the locus from the origin to the oil sample point, that is, Fn[X(O)/Z(O)] (Elzagzoug et al., 2014). Validation of this chromatic approach has been proven by comparing Fn[X(O)/Z(O)] values with colour index test results (Elzagzoug et al., 2014) (Figure 7.5).

An indication of trends in short-wavelength optical absorption (A) and long-wavelength light scattering (Sc) have been derived from the optical transmission results (R, G, B) (Sufian and Jones, 2017)

$$A = [1 - (Bt/Bo)/Gt/Go]$$

$$Sc = 2\{[(Rt/Ro)/(Gt/Go)] - 1\}$$

Both parameters have been shown to increase with the oil degradation level (Sufian and Jones, 2017) (Figure 7.6).

7.2.3 Summary

A chromatic system based upon a VDU illumination source and a mobile phone camera has been shown to be capable of monitoring optical transmission changes due to the degradation of transformer oils. The transmission results have been further analysed to provide chromatic insight into short-wavelength optical absorption and long-wavelength scattering which gives an indication of the presence of particles and sludge in the oil.

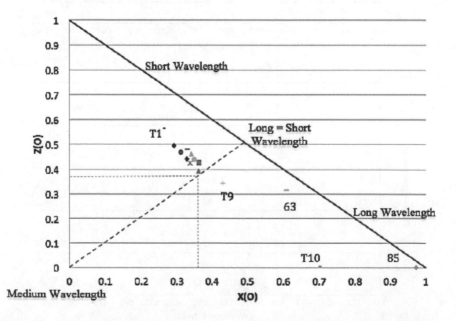

FIGURE 7.4   X, Y, Z Cartesian chromatic map for 14 oil samples from in-service transformers [(Z(O), X(O) are normalised Z, X values] (Elzagzoug et al., 2014).

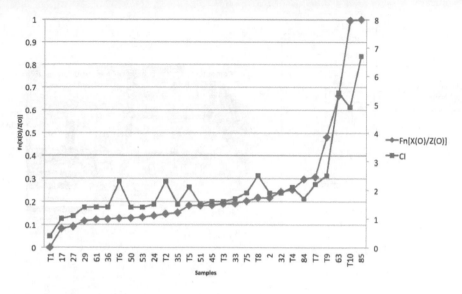

FIGURE 7.5 Comparison of chromatic parameter (Fn[X(O)/Z(O)]) with colour index results for 14 oil samples (Elzagzoug et al., 2014).

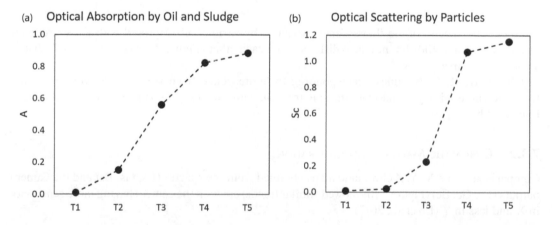

FIGURE 7.6 Variation of transmission-derived parameters with oil degradation (Sufian and Jones, 2017). (a) Short wavelength (absorption) (A); (b) long wavelength (scattering) (Sc). $A = [1 - (Bt/Bo)/Gt/Go]$; $Sc = 2\{[(Rt/Ro)/(Gt/Go)] - 1\}$.

7.3 CHROMATIC OPTICAL FLUORESCENCE

7.3.1 SYSTEM STRUCTURE AND OUTPUTS

An optical system which has been used for monitoring fluorescence produced by oil samples (Sufian and Jones, 2017) consisted of a ultraviolet (UV) laser light source (405 nm, 10 mW), addressing an oil sample contained in a cuvette and addressed with either a spectrometer or electronic camera whose output was fed into a data processing computer (Figure 7.7).

The options for the camera operation and so on were based upon the IMPS diagram (Chapter 2) (Figure 7.8).

The fluorescent emission from the oil sample was captured in a direction orthogonal to the transmission path of the exiting laser beam. Examples of images obtained with a camera are shown in Figure 7.9a, which shows the differences between water and clean and degraded transformer oil

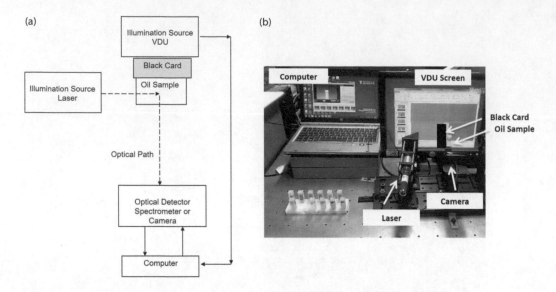

FIGURE 7.7 Optical fluorescence through transformer oils (Sufian and Jones, 2017). (a) Schematic diagram of general system structure for chromatic optical fluorescence monitoring; (b) view of the chromatic fluorescence oil monitoring system.

samples. The spectra of the fluorescence produced by various oils captured with a spectrometer (wavelength range 400–700 nm) also differed from each other (Figure 7.9b) (Lo et al., 2017), but in a complex manner.

Chromatic R, G, B values corresponding to fluorescence from various oils were extracted from the camera images and from the spectra – the latter using R, G, B processors, as shown in Figure 7.9b.

7.3.2 Chromatic Mapping and Calibration

Comparison of the X, Y, Z chromatic maps obtained from the spectra (Rs, Gs, Bs) and the camera outputs (Rc, Gc, Bc) (Figure 7.10) showed similar trends, but with the camera having more sensitivity in X and less in Y (Lo et al., 2017).

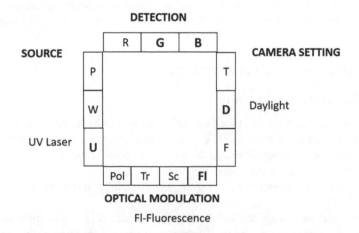

FIGURE 7.8 IMPS permutation of chromatic component properties for oil fluorescence tests (chosen operational parameters highlighted).

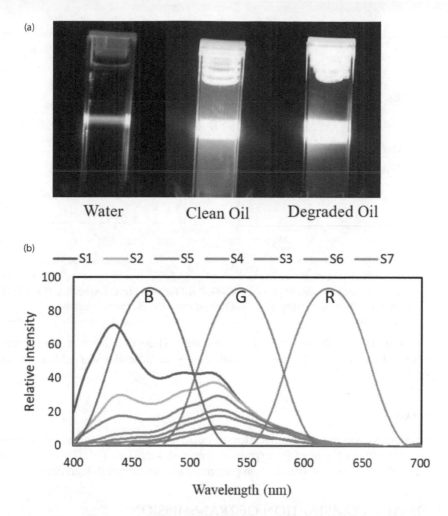

FIGURE 7.9 Oil fluorescence excited via UV laser beam (405 nm) (Lo et al., 2017). (a) Camera images (water, clean oil, degraded oil). (b) Fluorescence emission spectra for seven transformer oils (S1–S7) and R, G, B processor profiles.

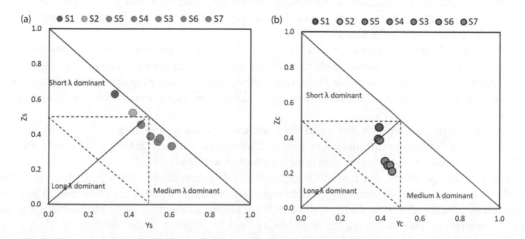

FIGURE 7.10 Chromatic X, Y, Z fluorescence maps for seven transformer oils (S1–S7) (Lo et al., 2017). (a) Derived from Rs, Gs, Bs of optical spectra. (b) Derived from Rc, Gc, Bc camera outputs.

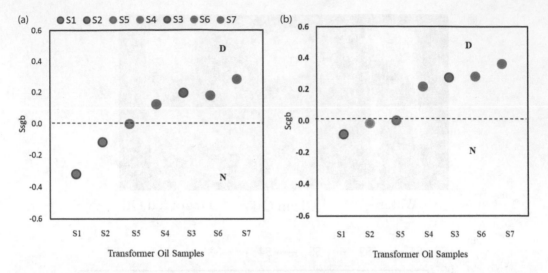

FIGURE 7.11 Chromatic nominal spread [S = (G – B)/(G + B)] for seven transformer oils (S1–S7) (Lo et al., 2017). (a) Derived from Gs, Bs of optical spectra. (b) Derived from Gc, Bc camera outputs.

The chromatic spread (S = (G – B)/(G + B)) varied with oil degradation in a similar manner for both the camera and spectra data (Figure 7.11), indicating S has potential as a calibration parameter (Lo et al., 2017).

7.3.3 SUMMARY

The chromatic approach was used with an electronic camera for monitoring laser-induced fluorescence in transformer oils without the need for expensive spectrometers (Lo et al., 2017). The chromatic spread parameter (S) was used as a calibration parameter for detecting oil degradation.

7.4 CHROMATIC COMBINATION OF TRANSMISSION AND FLUORESCENCE PARAMETERS

A compact, portable unit Optical Chromatic Transformer Oil Monitoring (OCTOM) unit has been developed which combines addressing optical absorption, scattering and fluorescence by an oil sample and operates automatically, processes the data chromatically and transmits the results wirelessly to a central hub (Sufian et al., 2018).

The optical absorption, scattering system structure (Figure 7.1a) and fluorescence system structure (Figure 7.7a) are combined into a single unit (Figure 7.12) for automatic operation. Illumination for absorption and scattering is produced by a miniature 4-inch visual display unit to provide tuneable, polychromatic light, whilst a monochromatic light beam for fluorescence is provided by an ultraviolet laser diode.

The optical outputs are captured by an electronic camera, and a miniature liquid crystal dis[play (LCD) screen displays preliminary test results. The system is controlled by a Raspberry Pi unit (Figure 7.12b), which also enables data to be transmitted wirelessly.

A typical image is shown in Figure 7.13 of an oil-filled cuvette illuminated simultaneously by the VDU for absorption and scattering evaluation and by the laser for fluorescence evaluation. The main areas in this image addressed for analysis were St for absorption/scattering, Sf for fluorescence and Ref for reference. Also shown are the three black points for image orientation checking.

The monitoring unit extracts R, G, B values from the transmission and fluorescence images and transforms them into reliable chromatic parameters (A, Sc, Flo; Sections 7.2.2 and 7.3.2) for

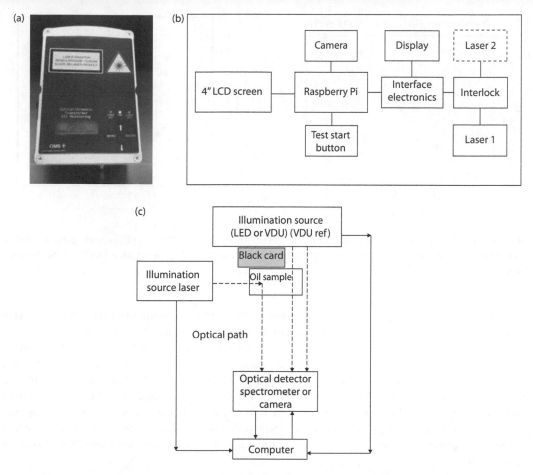

FIGURE 7.12 Combined transmission, scattering and fluorescence chromatic monitoring unit (OCTOM) (Sufian et al., 2018). (a) View of unit; (b) block diagram of the overall system; (c) schematic diagram of the optical layout of the combined transmission and fluorescence chromatic system (OCTOM).

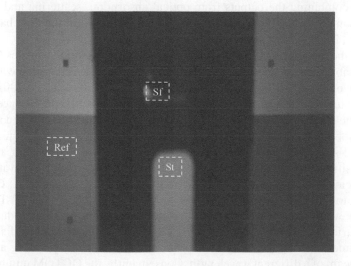

FIGURE 7.13 Location of reference and oil sample areas on images for analysis (Sufian and Jones, 2017). (St = oil sample transmission, Sf = oil sample fluorescence, Ref = reference area.)

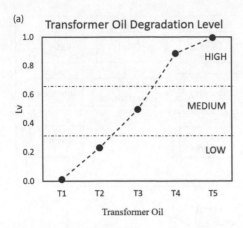

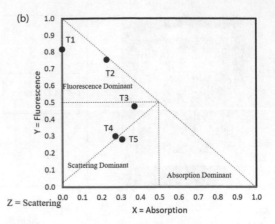

FIGURE 7.14 Overall chromatic calibration graphs (Sufian and Jones, 2017). (a) Chromatic oil degradation level (Lv versus transformer oil). (b) XYZ secondary chromatic map [X = Absorption (A), Y = Fluorescence (Flo), Z = Scattering (Sc)].

producing calibration graphs. The chromatic parameters are each normalised empirically to have a range of 0–1. A, Sc, Flo are then treated as three chromatic outputs (R, G, B) to produce X, Y, Z chromatic map parameters with a chromatic effective strength parameter [Lv = (A + SC + Flo)/3] which is indicative of the overall level of oil degradation. An empirically determined calibration graph of Lv versus degradation level is given in Figure 7.14a, which shows that normal oil samples are indicated by Lv < 0.33 and highly degraded oil samples correspond to Lv > 0.66, whilst moderately degraded oil samples lie in the range 0.33 < Lv < 0.66. An indication of which optical property dominated any degradation [Absorption (A), Scattering (Sc), Fluorescence (Flo)] is given by the coordinates of an oil on the X, Y, Z chromatic map, an example of which is shown in Figure 7.14b.

7.5 SUMMARY AND OVERVIEW

The process for combining the outputs from optical absorption, scattering and fluorescence for transformer oil samples using the portable OCTOM unit is shown in the flowchart in Figure 7.15 (Sufian and Jones, 2017).

Evaluation of the automated chromatic OCTOM unit with 81 real oil samples (TNB, Malaysia) confirmed its capability for distinguishing degraded oil samples when compared with Duval analysis (Singh and Bandyopadhyay, 2010), dissolved gases analysis (Guardado et al., 2001), acidity and so on. The OCTOM unit had an accuracy of 80% and a confidence level of high concern of 90% when compared with the combined predictions of Duval and other test methods (Sufian and Jones, 2019).

Further developments of the technique are possible. For example, the system response may be varied by utilising different camera settings (IMPS; Figure 7.8) (Sufian and Jones, 2017) to optimise sensitivity for particular types of oils and for optimising separately with regard to absorption, scattering and fluorescence. In addition, the raw values of R, G, B rather than the camera's automatically processed values may be acquired (A. A. Al-Tememy, private communication), whilst the chromaticity of the transmission illumination produced by the VDU screen can be varied via the OCTOM software. Fluorescence detection could also be varied by changing the camera setting (e.g., daylight to fluorescent) and incorporating a second laser to provide a laser beam of a different wavelength. Consequently, the OCTOM unit has a high degree of versatility.

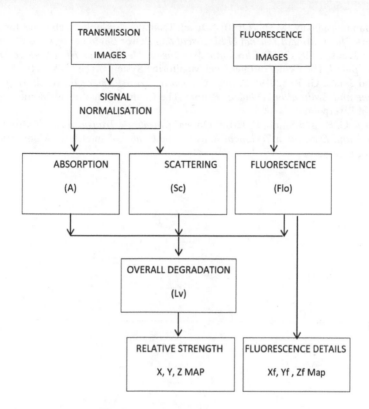

FIGURE 7.15 Chromatic combination decision chart for transformer oil samples using optical transmission and fluorescence (Sufian and Jones, 2017).

ACKNOWLEDGEMENTS

The provision of oil samples by Tenaga National Berhad Research (TNBR) in Malaysia and Electricity North West Ltd (ENW) in the UK is acknowledged.

REFERENCES

Bureau of Reclamation, U.S. Department of the Interior, Tech Svcs Group Hydroelectric Research. (April 2005) "Transformers: Basics, Maintenance and Diagnostics."

Elzagzoug, E. (2013) "Chromatic monitoring of transformer oil condition using CCD camera technology." PhD Thesis, University of Liverpool.

Elzagzoug, E., Jones, G.R., Deakin, A.G., and Spencer, J.W. (April 2014) "Condition Monitoring of High Voltage Transformer Oils Using Optical Chromaticity." *Institute of Physics, Measurement Science and Technology*, 25(6), 9.

Guardado, J. L., Naredo, J. L., Moreno, P., and Fuerte, C. R. (2001) "A Comparative Study of Neural Network Efficiency in Power Transformers Diagnosis Using Dissolved Gas Analysis." *IEEE Trans. Power Delivery*. 16(4), 643–647.

Lo, C. K., Looe, H. M., Sufian, A. T., Jones, G. R., and Spencer, J. W. (2017) "Transformer Oil Degradation Monitoring with Chromatically Analysed Optical Fluorescence." *Proceedings of the International Conference on Imaging, Signal Processing and Communication (ICISPC 2017)*. Association for Computing Machinery, New York, NY, USA, 171–175. doi: 10.1145/3132300.3132329

Lorin, P. (April-May 2005) "Forever Young [Longer-Lasting Transformers]." *Power Eng*, 19(2), 18–21, doi: 10.1049/pe:20050203

Nadkarni, R. A. (2007) *Guide to ASTM Test Methods for the Analysis of Petroleum Products and Lubricants*, vol 44 (Philadelphia, PA: ASTM International).

Singh, S., and Bandyopadhyay, M. N. (2010) "Duval Triangle: A Noble Technique for DGA in Power Transformers." *International Journal of Electrical and Power Engineering*, 4(3), 193–197.

Sufian, A. T., and Jones, G. R. (2017) *Chromatic Monitoring Methods for High Voltage Transformer Oils Technical Report 1*. The Center of Intelligent Monitoring Systems, The University of Liverpool.

Sufian, A. T., and Jones, G. R. (2019) *Optical Chromatic Transformer Oil Monitoring (OCTOM) Unit Performance and Evaluation Technical Report*. The Center of Intelligent Monitoring Systems, The University of Liverpool.

Sufian, A. T., Jones, G. R., and Smith, D. (2018) *Optical Chromatic Transformer Oil Monitoring (OCTOM) Performance and Evaluation Technical Report*. The Center of Intelligent Monitoring Systems, The University of Liverpool.

Section III

Chromatic Monitoring of Mechanical Vibrations

This section describes the use of the chromatic approach for extracting information about mechanical vibrations and acoustic signals produced by various sources. Examples of such deployments are the monitoring of cutting forces during mechanical machining of metallic materials, the live monitoring of running railway tracks from a moving, in-service train and monitoring the condition of high-voltage electric power equipment via acoustic signals from a transformer.

8 Mechanical Machining Monitoring

C. Garza

CONTENTS

8.1 INTRODUCTION

Monitoring cutting forces during mechanical machining is needed for reducing geometric errors, improving efficiency and optimising and modelling the processing (Sukvittayawong and Inasak 1991). Methods which have been used include the deployment of strain gauges, piezoelectric devices and conductive polymer and magneto-elastic materials (Elbestawi 1999). Efforts have been made to integrate force sensors into machine tools (Byrne and O'Donell 2007). Piezoelectric force transducers are susceptible to electromagnetic interference and do not have the dynamic response needed for high-speed machining. Optical methods such as the use of optical fibres and photo-elasticity have been proposed (Jina et al. 1995).

The use of optical chromatic monitoring of a photo-elastic element with optical fibre transmission of polychromatic light for such machine cutting is described in this section. Both static and dynamic machine operation are addressed. The machining of a mechanically rotated cylindrical rod carrying a rectangular slot along its length as a defect under real machining cutting conditions is used to assess the transient response, and comparisons are made with results obtained with a piezoelectric load cell.

8.2 TEST SYSTEM

An experimental system for evaluating the use of a chromatic optical method for monitoring machine cutting consisted of two parts – a mechanical cutting system and the monitoring system (Figure 8.1) (Garza 2010, Garza et al. 2012).

The mechanical cutting system consisted of a tool holder attached to a tool fixture via two screws (A, B; Figure 8.1). Static tests could be implemented via various degrees of compression obtained by adjusting one of the screws. For dynamic tests, a lathe was used with cylindrical rods, each carrying a 2-mm-wide groove and made from different materials (aluminium, brass, stainless steel). The rods were rotated at speeds up to 850 rpm.

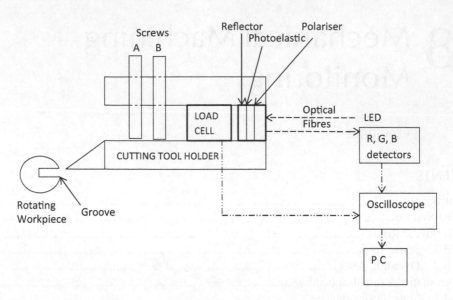

FIGURE 8.1 Chromatic system for monitoring the operation of a cutting tool.

The optical monitoring system was based upon a photo-elastic sensing element mounted between the tool fixture and cutting tool holder (Figure 8.1). The optical sensor consisted of a 3-mm-thick photo-elastic element with a reflective back plus polarisation filter addressed by white light from a light emitting diode (LED) source via a bundle of optical fibres, and the output was detected by three photo-detectors with overlapping wavelength responses. The optical fibre bundle consisted of seven, 400 mm to 400 μm fibres, with six fibres delivering the light to the sensor and one central fibre receiving the modulated light for transmission to the photo-detector. A piezoelectric load cell was attached to the tool cutter to provide comparative measurements of the stressed tool holder.

8.3 CHROMATIC RESULTS

Both static and dynamic chromatic test results have been reported (Garza 2010, Garza et al. 2012). The static results were obtained by applying a series of mechanical stresses to the tool cutter, whilst the dynamic results were obtained by rotating the cylindrical rod containing grooves on a lathe.

8.3.1 STATIC TESTS

A suitable location on a captured image of the photo-elastic element was chosen for addressing chromatically via the receiving optical fibre (Figure 8.1). The variation of each of the R, G, B output signals at this location with different mechanical loads (1–10 N) is given in Figure 8.2a, which shows how each of the three outputs varies differently with various loads. As a result, the values of the corresponding three chromatic parameters (H, L, S; Chapter 1) also tended to vary with the loads. An example of the variation of dominant wavelength H at the midpoint with a load is shown in Figure 8.2b. The variation is monotonic and so provides an unambiguous indication of the load value. A comparison of the susceptibility of the optical and piezoelectric orthogonal stresses indicated that the susceptibility of the optical sensor was <2% (Garza 2010, Garza et al. 2012).

8.3.2 DYNAMIC TESTS

The response of the photo-elastic sensing system (Figure 8.1) has been shown to be independent of signal frequencies up to 7.5 kHz and within 2.2% for frequencies up to 15 kHz (Garza 2010,

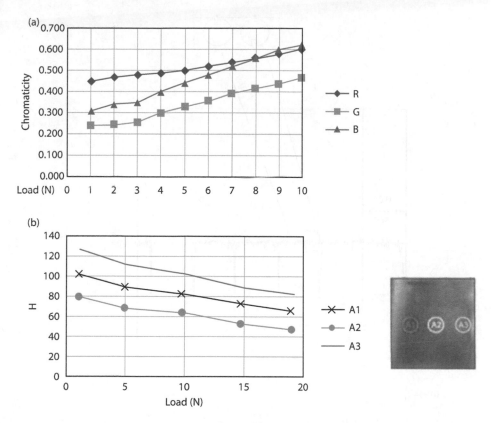

FIGURE 8.2 Static chromatic addressing of the photo-elastic element. (a) Example of R, G, B outputs as a function of mechanical load. (b) Examples of chromatic H at three locations as a function of mechanical load.

Garza et al. 2012). The time response of the entire system (Figure 8.1) with optimised photo-elastic sensing and using the chromatic H parameter are shown in Figure 8.3a and b. The variation of the H parameter with time for turning of an aluminium rod at 850 rpm with major pulses occurring every 0.07 s is shown in Figure 8.3a. A time-expanded view of both chromatic H and piezoelectric output (Lc) during a groove transit with aluminium at 850 rpm is shown in Figure 8.3b.

8.4 CHROMATIC RESULTS IMPLICATIONS

The test results indicated that the chromatic approach can address the complex structure of the optical pattern produced by a photo-elastic element subjected to machine operating conditions. Consequently, chromatic parameters H, L, S can in principle be used for monitoring such machine cutting operations. The H parameter provides a first-order quantification with high optical efficiency compared with monochromatic detection and with less susceptibility to spurious monochromatic disturbances. Also, chromatic H and S are independent of optical intensity and so provide some immunity to extraneous light intensity variations.

8.4.1 STATIC PERFORMANCE

Under static conditions (Figure 8.2), the chromatic parameter H varies monotonically over a range of 55° for a load change of 0–10 N with a coefficient of regression $R^2 = 0.99$. A sensitivity of 6°/N with a noise level of only 1° and a resolution of 0.164 N have been demonstrated. A sensitivity to orthogonal stresses was <2%, with a repeatability of <2°.

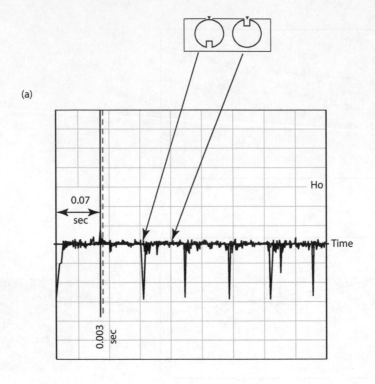

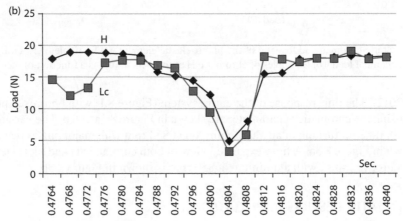

FIGURE 8.3 Dynamic test results for a cylindrical rod with groove. (a) Time variation of chromatic parameter H. (b) Fine time scale variation of chromatic H and piezoelectric output (Lc).

8.4.2 DYNAMIC PERFORMANCE

The dynamic performance of the chromatic photo-elastic unit was evaluated via two graphs (Garza 2010, Garza et al. 2012). The first was the use of the chromatic parameter H values to determine the rotational period of the machined rods (i.e., time between pulses in Figure 8.3a) versus the nominal rotation rate (Figure 8.4a). The second was the use of the chromatic parameter H values to determine the pulse width versus the calculated value from the nominal rotation speed and known groove width (Figure 8.4b).

The chromatic photo-detection unit had no change in attenuation for frequencies below 7.5 kHz but had a change in attenuation of 2.2% for frequencies above 7.5 kHz up to 15 kHz. The angular speeds of rotation of the grooved rod deduced from the time separation between consecutive

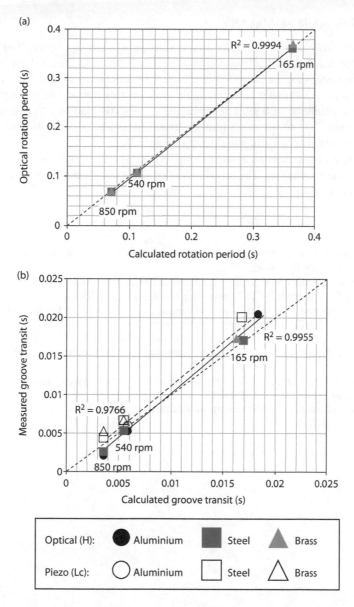

FIGURE 8.4 Measured versus calculated time scales for rotating workpieces of different materials and rotation rates: (a) rotation period from chromatic H measurements; (b) groove transit time from chromatic H and piezoelectric load cell output.

chromatic H pulses (Figure 8.3a) showed a linear variation with the nominal rotation rate of the rod for the rotational speed range of 165–850 rpm (Figure 8.4a). The linear variation was similar for the different rod materials (aluminium, steel, brass), and the sensitivity was 9.4°/N.

The values of the pulse width associated with the rotating groove determined chromatically showed good agreement with the calculated values (Figure 8.4b) for the same rotation rates (165–850 rpm) and the different rod materials (aluminium, steel, brass). Comparison of the piezoelectric sensor results for the pulse width showed a greater scatter than the chromatic sensor results (Figure 8.4b).

The coefficient of determination (R^2) for the chromatically measured and calculated periods of rotation (Figure 8.4a) was 0.999, whilst a comparison of the groove transit time (Figure 8.4b) gave $R^2 = 0.996$. The piezoelectric measurements of groove width (Figure 8.4b) gave $R^2 = 0.977$.

8.5 OVERVIEW AND SUMMARY

The feasibility of using a chromatically based sensing system for addressing a photo-elastic element stressed by the mechanical rotation of a mechanically machined cylindrical rod has been demonstrated. The system did not use mono-mode optical fibres nor monochromatic light. Calibration graphs of the chromatic parameter H versus static load, turning period and groove transit time had coefficients of determination of 0.99. The repeatability of static load results from repeated tests is claimed to be at least 2%, whilst machining different materials agreed with each other to within +/−1.5%, and susceptibility to orthogonal stresses was <2%.

Further investigations are needed to assess long-term drift effects and susceptibility of the system to ambient temperature variations. The possibility of chromatically extending the force measurement range also warrants investigation.

REFERENCES

Byrne, G. and O'Donell, G. E. (2007) An integrated force sensor solution for process monitoring of drilling operations, *Ann. CIRP* 56, 89–92.

Elbestawi, M. A. (1999) Force measurement in *The Measurement, Instrumentation and Sensors Handbook* Chapter 23 (J. G. Webster Ed.) CRC Press Boca Raton, FL, USA, 23-7–23-9.

Garza, C. (2010) Investigation of an optical sensor for cutting force measurement through chromatic modulation, PhD Thesis, University of Liverpool.

Garza, C., Jones, G. R., Hon, K. K. B., Deakin, A. G. and Spencer, J. W. (2012) Feasibility of monitoring mechanical machining with a chromatically addressed optical fibre photo-elastic sensor, *Strain*, 49, 68–74.

Jina, W.L., Venuvinod, P. K. and Wang, X. (1995) An optical fibre sensor based cutting force measuring device, *Int. J. Mach. Tools Manufact*, 35, 877–883.

Sukvittayawong, S. and Inasak, I. (1991) optimisation of turning process by cutting force measurement, *ISME Int. J. Ser. C*, 34, 546–555.

9 Chromasonics
Monitoring Rail-Track Faults

R. K. Todd

CONTENTS

9.1 INTRODUCTION

Railway operators have a need to ensure that their running track is in good condition and safe for freight and passenger traffic. Current methods for doing so include video scanning of the rails, eddy current generation, ultrasonics and walking the tracks.

The first two methods are unable to detect many rail faults, especially in the early stages. The latter is slow and can only detect obvious effects. Ultrasonic methods have been shown to perform well (Hesse, 2007) in detecting nearly all rail defects but are expensive in requiring an ultrasonic test car and involve slow operations (about 40 mph) along the tested rail track. This means that they can only be used when main line traffic is not running, Thus, a method based upon the use of a high-speed main-line locomotive which provided fault detection during normal operation would be advantageous. Such a possibility arises from the use of chromatic techniques in the acoustical domain called Chromasonics. Such a system fitted on a locomotive would be cost effective compared to ultrasonics, could be used daily for monitoring during normal running schedules and could relay information to a control centre whence remedial action could be taken. Chromasonics has been tested and run in the United Kingdom and United States and shown to be capable of detecting many important rail defects. An example is shown in Figure 9.1.

A description of the Chromasonics approach is given along with examples of defects detected in real tests with the method.

9.2 CHROMASONICS

Chromasonics is the adaptation of chromatic techniques in the acoustical domain for monitoring acoustic signals from railway tracks on board a locomotive during its passage along the railway track. The basis of the Chromasonics approach is described, followed by a description of instrumentation used for its deployment on board operating locomotives.

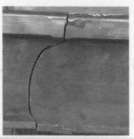

FIGURE 9.1   Vertical crack on a rail track.

9.2.1 Principles of Operation

The adaptation of chromatic techniques for monitoring acoustic signals involves deploying three acoustic filters (R, G, B) with non-orthogonal frequency responses to address a Fourier-transformed acoustic signal in the frequency domain. This compares with deployment of three non-orthogonal optical filters for processing optical signals as described in Chapter 1.

Figure 9.2 shows three frequency domain acoustical signals at each of three different times, the first corresponding to a normal operating condition and the other two to two different fault conditions at later times. Also shown in Figure 9.2 are the responses of three non-orthogonal acoustic filters (R, G, B) superimposed upon the three acoustic signals in the frequency domain.

The outputs from the three acoustic filters are chromatically processed to yield three chromatic parameters H, L, S (Chapter 1) which represent the dominant frequency, effective signal strength and signal spread and which can be represented on polar diagrams of H:S and H:L (Chapter 1). Figure 9.3a and b show typical representations of the H, L, S signature of an acoustic signal. Thus, different acoustic signals may be graphically represented by their coordinates on such H:S and H:L polar maps.

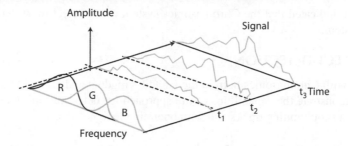

FIGURE 9.2 Chromatic processors R, G, B addressing acoustical signals at different times (t_1, t_2, t_3).

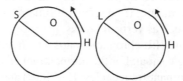

FIGURE 9.3 Schematic chromatic polar diagrams for a signal O at a given time: (a) H:S, (b) H:L.

9.2.2 MONITORING SYSTEM

The basic structure of a Chromasonics system is shown in Figure 9.4. This consists of a standard microphone in association with a personal computer (PC) and data storage means plus data transfer capability.

A microphone with a frequency range of 10 Hz to 20 kHz was mounted on the locomotive in a suitable position for receiving acoustic signals from the rail track produced by the rotating wheels of the locomotive. The output from the microphone was fed into a PC for chromatic processing. The outputs were Fourier-transformed into frequency domain signals for chromatic monitoring. The resulting information was stored on the unit with a provision for wireless transmission to a central control hub via general purpose radio system.

A chromatic system has been tested in the United Kingdom and United States. Extensive testing with Union Pacific Railways USA enabled comparisons to be made with results from an ultrasonic

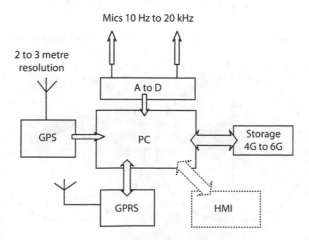

FIGURE 9.4 Chromasonics data collection and processing system.

test vehicle which indicated that the Chromasonics system could detect most defects indicated by the ultrasonic system.

9.3 RAIL-DEFECT DETECTION

Tests performed with Chromasonics both in the United Kingdom and United States have provided results which demonstrate the capabilities of the approach for detecting rail defects with H:S and H:L chromatic maps on running tracks and with operating locomotives.

9.3.1 NORMAL TRACK SIGNAL PLUS SQUAT

Squats are an indentation in the rail head which tend to fill up with dirt and grit and are black in appearance. They are frequently caused by track ballast on the rail track run over by a locomotive wheel. They are of variable size, ranging from a few millimetres to a few centimetres. A squat can lead to the formation of defects such as vertical cracks.

Figure 9.5a and b show H:S and H:L chromatic maps which display signals from both a normal length of railway track plus a single squat. For the normal track, the dominant frequency (H) varied continuously with mid-range dominant frequencies (H $\sim$ 180°) and a relatively low level of effective magnitude (L). However, a squat is clearly distinguishable with a high value of effective magnitude (L) and a well-defined frequency (H) of dominant value less than 90°.

9.3.2 NORMAL TRACK SIGNAL PLUS HORIZONTAL CRACK

Figures 9.6a and b show H:S and H:L chromatic maps which display signals from both a normal length of railway track plus a horizontal crack. The normal track has a different signature from the

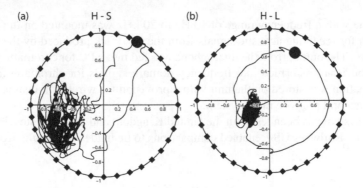

FIGURE 9.5 Chromatic polar diagrams of a normal railway track and squat signals: (a) H:S, (b) H:L.

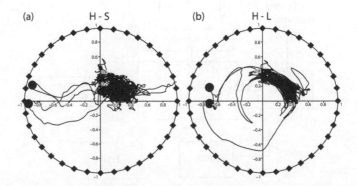

FIGURE 9.6 Chromatic polar diagrams of a normal railway track and crack signals: (a) H:S, (b) H:L.

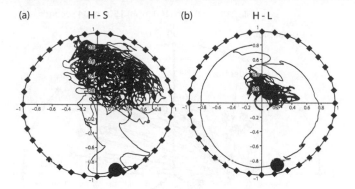

FIGURE 9.7   Chromatic polar diagrams of a normal railway track and defective weld signals: (a) H:S, (b) H:L.

track signal of the railway track shown in Figure 9.5 but with some similar features. The signal shows similar variability but has a lower spread (S) and a lower dominant frequency (H), with a relatively low effective strength (L).

The presence of a horizontal crack of about 15 cm length produced a narrow frequency band (S) signal with a much higher dominant frequency (H) than the background rail signal. Two points are apparent which may give an indication of the length of the crack.

9.3.3   Normal Track Signal Plus Defective Weld

Figures 9.7a and b show H:S and H:L chromatic maps which display signals from a normal length of railway track plus a defective weld between two lengths of the track. The normal track signal in this case showed substantial variation in signal frequency spread (S) and the dominant frequency (H) at relatively low levels of effective signal magnitude (L).

The defective weld produced a distinctive signal with a much higher dominant frequency (H) and narrower spread (S) plus a higher effective magnitude (L).

9.3.4   Worn Track Signatures

Figures 9.5–9.7 show that the chromatic signatures of lengths of different tracks can vary with regard to the dominant frequency (H) and spread (S) but less so for the effective magnitude (L). However, extensive tests on different tracks have shown that the acoustic signature of a railway track tended to vary as it became older (Figure 9.8). Nonetheless, defects were observed to be located in the same frequency domain on both H:S and H:L chromatic maps. The H:L map of Figure 9.8 shows how the presence of squats was detectable regardless of the track being new or old. Figure 9.8 also shows the occurrence of wheel burn (i.e., the locomotive drive wheels spinning without the locomotive moving due to a heavy load) and the track becoming excessively heated and deformed.

9.4   OVERVIEW AND SUMMARY

Real track tests with Chromasonics test units on board travelling locomotives indicate that the Chromasonics unit is capable of detecting defects on railway running tracks during normal locomotive operation (Todd, 2005, 2007).

The tests have shown that different tracks can have different H, L, S coordinate values which for each track form clusters together. The occurrence of various faults (squats, cracks, faulty welds) are identifiable by changes in the H, L, S signatures being distinctly different from the normal track signatures. Different faults may also be distinguishable from each other. However, further

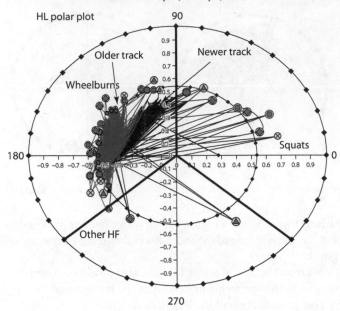

FIGURE 9.8   Chromatic H:L polar diagram of new and old railway tracks.

research and evaluation are needed to realise the full potential of the technique over a wide variety of conditions.

REFERENCES

Hesse, D. (2007) Rail Inspection Using Ultrasonic Surface Waves, Department of Mechanical Engineering Imperial College London SW7 2AZ, (pdf Section 3 P. 43)

Todd, R. K. (2005) Rail Flaw Detection using Acoustic Techniques for Network Rail (File reference PRJ-DF-LD 79565; Page 46; Section 7.5; Chromasonic Data analysis and appendix A and B)

Todd, R. K. (2007) Chromatic processing technology: Defect detection of railway tracks. doi: 10.1049/etr.2016.0090; ISSN 2096-4007; www.ietdl.org

Section IV

Environmental Applications

The use of chromatic techniques has been reported for monitoring biodigestion at a waste-recycling unit and airborne microparticle pollution in urban areas (Jones et al., 2008). The chromatic approach has been extended for other environmental problems. These include an extension of microparticle monitoring for long free path applications, monitoring marine water samples, evaluating the use of electric power generation by large-scale windmills and assessing the environmental suitability of various gases used for the electrical insulation of high-voltage electric power equipment. These applications use different forms of chromaticity – optical absorption/scattering for long path particle monitoring and marine water assessment, time-based analysis of windmill power production and severity of air-quality reduction potential caused by different electrical insulating gases. Details of chromatic deployment for these various environment-related aspects are considered in this section.

Section IV

Environmental Applications

10 Optical Chromatic Monitoring of Marine Waters

J. L. Kenny

CONTENTS

10.1 INTRODUCTION

Measuring the inherent optical properties (IOPs) of sea water from real environments (Figure 10.1) is important for understanding the propagation of light through water to determine how much light is available for primary production (photosynthesis) and for optical communications with low-energy communications between underwater autonomous vehicles. Ocean colour remote sensing models and measurements rely on a knowledge of the optical backscatter from particulate matter in the water, which can also provide information about the particulate matter characteristics (e.g., bulk refractive index, particle size distribution and composition) (Korotaev and Baratange, 2004), which is essential to understanding the optical variability of natural waters.

As instrumentation for measuring the inherent optical properties of marine waters becomes more advanced, the data produced by the emerging instruments (e.g., Wetlabs AC-S) is becoming more complex. Where simple graphical representations may have been sufficient for reductionist data (e.g., the backscattering coefficient at a single wavelength), this is not the case for complex data. Chromatic methodology offers a means for balancing the range of relevant information with simplicity of presentation that allows for easy assimilation of information by non-specialist personnel or autonomous systems. The usefulness of the chromatic approach has been demonstrated with real marine water tests from sites 1, 2 and 3 of the river estuary and a fourth sample taken from the beach marked Hilbrie Island shown in Figure 10.1.

10.2 INHERENT OPTICAL PROPERTIES OF WATER

Natural water has two fundamental inherent optical properties

1. The absorption coefficient (a_c) defined by (absorbed incident power)/(optical path length)
2. The volume scattering function (s_f) defined by [(light intensity 0°–180°)/(incident irradiance/volume)]

A total scattering coefficient is defined as the integral of s_f from 0° to 180°, whilst a backscattering coefficient is given by the integral of s_f from 90° to 180° (Mobley, 1994).

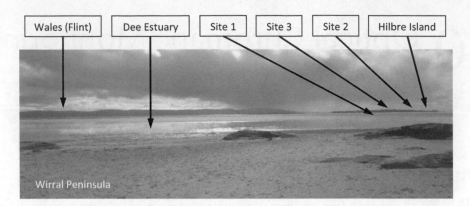

FIGURE 10.1 Example of river estuary from which marine waters were monitored (Hilbre Island, Dee Estuary. Merseyside, UK [facing south]).

The amount of absorption and scattering of light in natural waters is dependent not only upon the absorption and scattering coefficients of the water but also on the optical properties of material dissolved or suspended within the water. These include Raman scattering by water molecules, fluorescence by coloured dissolved organic matter (CDOM), absorption, scattering and fluorescence by organic material (phytoplankton and algae) and absorption and scattering by non-organic material (sand, detritus and sediments).

10.3 CHROMATIC PROCESSING

Polychromatic optical signals from light transmission through complex marine waters may be addressed with chromatic techniques (Chapter 1). A preferred chromatic approach has been the use of the Lab method. The algorithms of the Lab method are defined by the following equations (Schwarz et al., 1987; Ainouz et al., 2006):

$$L = 116 \ (G/Gn)^{1/3} - 16) \tag{10.1}$$

$$a = 500 \ [(R/Rn)^{1/3} - (G/Gn)^{1/3}] \tag{10.2}$$

$$b = 200 \ [(G/Gn)^{1/3} - (B/Bn)^{1/3}] \tag{10.3}$$

for R/Rn, G/Gn, B/Bn > 0.008856
where Rn, Gn, Bn are processor outputs when addressing data to be used as a reference source.

$$L = 0 \text{ (black) to } 100 \text{ (white)}; \quad Hab = \tan^{(-1)} (b/a); \quad Sab = (a^2 + B^2)^{1/2}$$

A reference signal is used to calculate the chromatic values; lightness (L), A and B, where L is a representative signal strength, "a" represents the chromatic space from red (+a) to green (−b) and "b" represents the chromatic space from yellow (+b) to blue (−b).

10.4 CHROMATIC MONITORING SYSTEM

An optical system for monitoring marine water (Figure 10.2) has been developed at the National Oceanography Centre in Liverpool, UK, as part of Oceans 2025, the Natural Environment Research Council (NERC)'s proposed strategic research programme. This uses a dual halogen/deuterium light source to provide broadband white light focused onto a linear variable bandpass optical filter

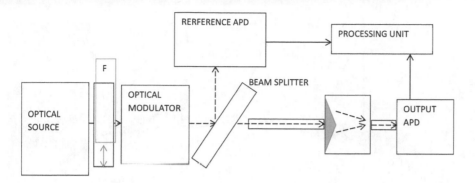

FIGURE 10.2   Schematic diagram of the optical monitoring system.

to provide a narrow optical bandwidth of approximately 40 nm. As the filter is displaced, the centre optical wavelength is shifted along the spectral range of the source. The light then passes through an optical modulator operating at 200 Hz with a 90% duty cycle, which is then divided by a glass plate, angled at 45° from the normal (Figure 10.2). Part of the light is reflected onto an avalanche photo diode (APD) to provide a reference to compensate for variance in the source output. The remaining light is focused onto the face of a 1 mm diameter optical fibre through which the light is transmitted into a water sample before being transmitted via another optical fibre to a second avalanche photo diode.

The volume scattering function at a fixed angle was calculated by multiplying the sensor response (normalised against a reference signal representing the output from the light source) by the sensitivity coefficient and the attenuation compensation function.

The sensitivity coefficient was derived by moving a plaque of spectralon (with a known reflectivity factor) away from the face of the sensor. The attenuation compensation function corrected for the attenuation of the light along the transmission and return path through the sample water. It is dependent on spectral attenuation measurements of the sample water (performed in this case by a Wetlabs AC-S).

The backscattering coefficient was then calculated using the methods described by (Maffione and Dana, 1997; Dana and Maffione, 2002).

The spectral range of the Wetlabs AC-S limited the "corrected" range of the optical system between 400 and 750 nm.

Some test results were obtained for grab samples of marine waters with natural sediment taken from four sites in the River Dee Estuary (Figure 10.1). The samples were tested with a test tower which circulated the water for testing with different systems, including the chromatic system represented in Figure 10.2.

10.5 EXAMPLES OF TEST RESULTS

An important monitoring parameter was the backscattering coefficient (Section 10.2) (Mobley, 1994; Kenny, 2015).

The corrected backscattering coefficients for water samples from the four sites (1–3 and Hilbre beach) on the Dee estuary (Figure 10.1) are shown as a function of optical wavelengths in Figure 10.3. This shows the complex variation of the marine water between these four sites. The highest overall backscattering occurred for the sample from the deeper channel (site 1), whilst the lowest was for the shale (site 3) on the island (Hilbre). The sample from the island with red shale (site 2) had a distinctive broad peak at about 600 nm wavelength.

This figure illustrates the complexity of the test results but also some clear differences between the four samples which are not easily quantified from the backscattering characteristics. However,

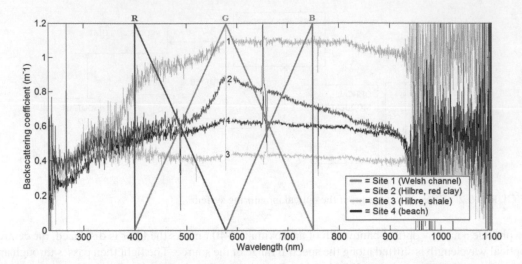

FIGURE 10.3 Corrected backscattering versus wavelength results using Sea Krait OBS monitoring system (Kenny, 2015).

chromatic Lab addressing (Section 10.3) methods may be deployed to quantify the different marine water characteristics from each of the four sites.

Triangular RGB receptors were applied to the data between 400 and 750 nm (as shown in Figure 10.3).

The reference values Rn, Gn, Bn were obtained by applying triangular RGB receptors to the backscattering coefficient for pure water (as presented by Morel, 1974) between 400 and 750 nm.

$$b_w(\lambda) = 1.458 * 10^{-3} \left(\frac{550}{\lambda} \right)^{4.34} \tag{10.4}$$

where $b_w(\lambda)$ is the backscattering coefficient of pure water and λ is the wavelength of the light.

Figure 10.4 shows the Lab plot calculated from the backscattering measurements shown in Figure 10.3. The results obtained for the four sites of Figure 10.1 are shown, from which the difference in the spectral signatures of the backscattering coefficient can be quantified in reference to the backscattering coefficient of pure water. This provides a convenient way for presenting the

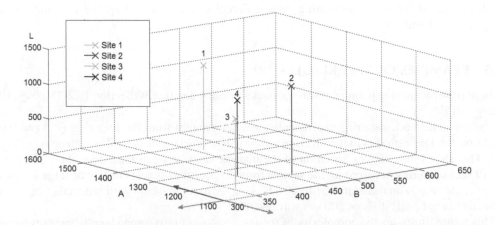

FIGURE 10.4 Chromatic Lab map of backscattering coefficient from Sea Krait OBS (Kenny, 2015).

information for an operator to observe and interpret. For example, the Lab graph shows that the sediment from site 2 scatters more blue light than the sediment from the other sites and that the sediment from site 1 scatters more light across the measured spectrum at the concentrations used than the sediment from the other sites.

10.6 SUMMARY AND OVERVIEW

The chromatic Lab diagram provides a useful categorisation means for various marine water samples. The approach could be used by either an automated system or as an input for a machine learning system. Further work could also provide additional information for marine water studies.

ACKNOWLEDGEMENTS

Special thanks to Prof. Joe Spencer (University of Liverpool) for all his advice, to Professor Alex Souza (National Oceanography Centre) and to Dr Mike Smithson (National Oceanography Centre) for all their support.

REFERENCES

Ainouz, S., Zallet, J., de Martini, A., and Collet, C. (2006). Physical interpretation of polarisation-encoded images by clour preview, *Opt. Express*, Vol. 14, No. 13, pp. 5916–5927.

Dana, D.R. and Maffione, R.A. (2002). Determining the back scattering coefficient with fixed-angle backscattering sensors - revisited. *Ocean Optics XVI*, Santa Fe New Mexico.

Kenny, J.L. (2015). Optical properties of marine waters – A method based on chromaticity. PhD Thesis, University of Liverpool, Liverpool, UK

Korotaev, G. and Baratange, F. (2004). Particulate backscattering ratio at LEO 15 and its use to study particle composition and distribution. *Journal of Geophysical Research*, Vol. 109, p. C01014.

Maffione, R.A. and Dana, D.R. (1997). Instruments and methods for measuring the back-scattering coefficient of ocean waters. *Applied Optics*, Vol. 36, No. 24, p. 6057.

Morel, A. (1974). Optical properties of pure water and pure sea water. *Optical Aspects of Oceanography*. Edited by N.G. Jerlov and E. Steemann Nielsen, pp. 1–24, Academic Press, San Diego, Calif.

Mobley, C.D. (1994). *Light and Water: Radiative Transfer in Natural Waters*. Academic Press, San Diego, Calif.

Schwarz, M.W., Cowan, W.B., and Beatty, J.C. (1987). An experimental comparison of RGB, YIQ, LAB, HSV, and opponent color models. *ACM Transactions on Graphics*, Vol. 6, No. 2, pp. 123–158.

11 Chromatic Analysis of Wind Power Generation

A. T. Sufian, J. Lawton and G. R. Jones

CONTENTS

11.1 INTRODUCTION

The generation of electric power from wind-driven generators (Figure 11.1) relies upon sufficient wind being available. In situations where there is an absence of sufficient wind, there is an intention to transfer power from another area where there is sufficient wind to produce power. For example, power interconnections between the United Kingdom and Germany allow power to be supplemented from one to another when one is short of wind but the other has sufficient wind. However, in practice, it needs to be ascertained how frequently both areas may have wind deficiencies which coincide with each other. Data about the power generation from wind at different times in the United Kingdom and Germany is available (Renewal Energy Foundation; Fraunhofer Institute) and has been analysed to provide a quantitative evaluation of such behaviour. The adaptation of chromatic analysis of such a situation is described.

11.2 WIND SPEED DATA

Wind speed data for the United Kingdom and Germany for an 8-month period (January and December 2015–2018) has been considered as a percentage of the maximum wind speed for each month and normalised to be in the range 0 (minimum) to 1 (maximum). Typical wind speed data from the United Kingdom (Renewable Energy Foundation) and Germany (Fraunhofer Institute) normalised in this way is shown in Figure 11.2 as a function of days over a period of 1 month (January 2015).

11.3 DATA ANALYSIS

11.3.1 CHROMATIC ANALYSIS

Wind speed data of the form shown in Figure 11.2 has been analysed by chromatically processing the data for each country using three non-orthogonal detectors in the time domain (R, G, B) with triangular responses to provide uniform sensitivity over the whole range being addressed (Chapter 1). The detector outputs are processed to yield values of chromatic parameters H, L, S, with H being the dominant time along the signal time period, L the effective strength of the signal over that period and

FIGURE 11.1   View of wind farm for electric power production.

S the spread (Chapter 1). The values of each of these parameters (H, L, S) for the UK and Germany wind power signals are compared on graphs of the UK versus Germany values (Figure 11.3) over a number of selected time periods.

The dominant time H $\rightarrow$ 0 corresponds to the early part of the signal, whereas H $\rightarrow$ 1 corresponds to the final part of the chosen time period. The effective strength L $\rightarrow$ 0 indicates no signal, whilst L $\rightarrow$ 1 is a maximum signal. The spread S $\rightarrow$ 0 indicates a wide spread, whilst S $\rightarrow$ 1 indicates no spread.

11.3.2   Statistical Analysis with Chromatic Parameters

The chromatic parameters H, L, S have been further processed to highlight various statistical features using a Bland and Altman plot (Bland and Altman, 1986). This is based upon the difference between the values of a chromatic parameter (H, L or S) between the United Kingdom and Germany [e.g., (UK)L − (Germany)L] versus the mean of the UK plus Germany values (e.g., [(UK)L + (Germany)L]/2]; Figure 11.4).

A mean difference (MD = (sum of [UK − Germany])/[number of data sets]) = 0 indicates no difference between the United Kingdom and Germany. The spread of the data from the mean difference yields a limit of agreement (LoA) indicating the standard deviation(s), which is a limit

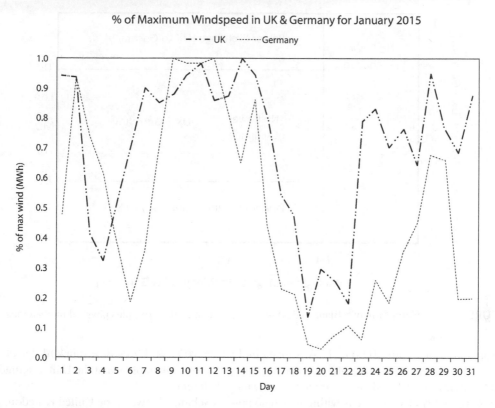

FIGURE 11.2 Result showing typical time variation of normalised wind speeds in the United Kingdom and Germany (January 2015) (Sufian et al., 2019).

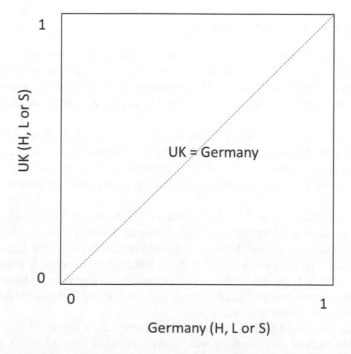

FIGURE 11.3 Graph comparing values of a chromatic parameter (H, L or S) of the United Kingdom and Germany (Sufian et al., 2019).

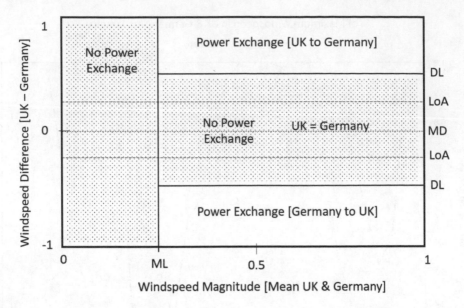

FIGURE 11.4   General form of a Bland and Altman plot (Sufian et al., 2019) plus physical implications.

of spread of data from MD (i.e., LoA is given by LoA = MD +/– 1.96 s [s = standard deviation]; Bland and Altman, 1986). An MD = 0 and a small spread of LoA implies that the United Kingdom and Germany wind speed values are not significantly different.

Figure 11.4 shows two areas within which no power exchange between the United Kingdom and Germany would occur due to either the wind speed difference (DL) or the wind speed magnitude (ML) being too low (Figure 11.4).

11.4   RESULTS OF TEST DATA ANALYSIS

The previous methods of data analysis have been applied to the chromatic parameters H, L, S for three different-duration time periods of an 8-month test period of wind power production in both the United Kingdom and Germany. The durations of the three time periods analysed were a month from each of the 8 months monitored, weekly for the 8 months and every 3 days for the 8 months.

Examples of the deployment of the chromatic signal processors (R, G, B) over a month of data for the three different time durations (month, week, 3 days) are shown in Figure 11.5a–c, respectively.

The Bland and Altman plots for the effective magnitude (L) corresponding to the three time durations (month, week, 3 days) extending over the 8-month test period are shown in Figure 11.6a–c, respectively.

Similar Bland and Altman plots have been obtained for the H and S chromatic parameters.

The monthly processed L data (Figure 11.6a) showed a small amount of power magnitude shifts over the 8-month test period. However, the differences between the United Kingdom and Germany were small. The LoA was less than 0.15, far less than the acceptable power exchange limit (DL = 0.5), although UK wind was slightly stronger than that in Germany, as indicated by MD > 0. The dominant time H data showed a similar trend, with the UK and Germany values being close, as were the trends with the spread of power S.

For the weekly processed L data (Figure 11.6b), there was a higher scatter of data, but the difference between the UK and Germany wind power remained low, with LoA less than 0.3 less than the acceptable power exchange limit DL. There were similar trends in the dominant time (H) and spread (S).

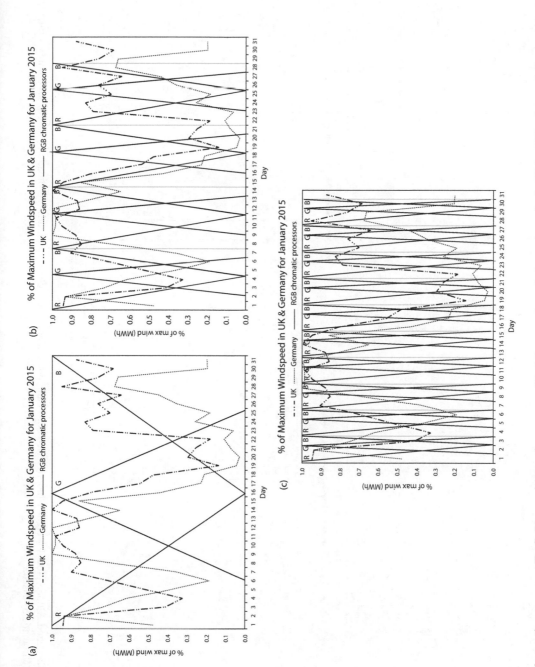

FIGURE 11.5 Chromatic processors (R, G, B) superimposed upon normalised time-varying wind power graphs for United Kingdom and Germany 1-month window (Sufian et al., 2019). (a) One-month duration; (b) 1-week duration; (c) 3-day duration.

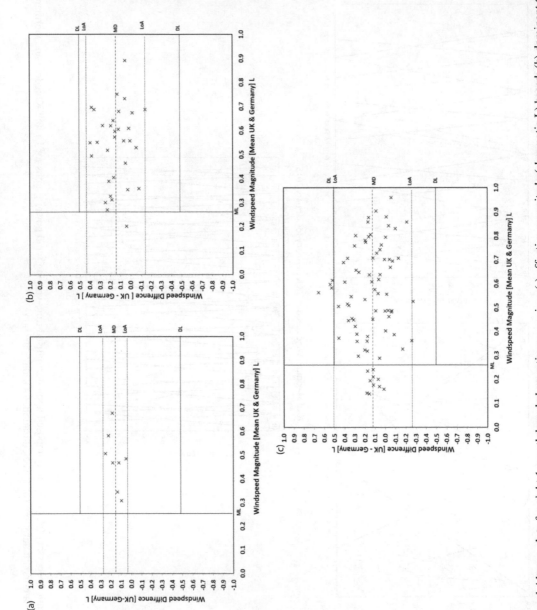

FIGURE 11.6 Bland and Altman plots for eight 1-month based chromatic processing: (a) effective magnitude (chromatic L) based; (b) dominant time (chromatic H) based; (c) spread (chromatic S) based.

The 3-day processed L data (Figure 11.6c) showed a variation in the effective magnitude and wind speed difference between the United Kingdom and Germany. However, the acceptable power exchange from the United Kingdom to Germany was only 5% of the data. The remaining L data were below the acceptable power exchange limit DL, with LoA below 0.4, indicating that no substantial power exchange between the United Kingdom and Germany was practicable. Similar trends were observed for the dominant time (H) and spread (S).

11.5 SUMMARY AND OVERVIEW

Chromatic analysis of the variation in power produced in the United Kingdom and Germany in combination with Bland and Altman graphs provided quantitative insight about addressing power shortages by transferring power between the two countries.

Chromatic processing combined with Bland and Altman plots provided a means for quantifying the likelihood of wind-generated electric power being transferred between the United Kingdom and Germany in the event of low wind occurrence in one country. A primary indication of whether there is an imbalance in the wind power between the United Kingdom and Germany is given by the chromatic effective magnitude parameter (L). The dominant time parameter (H) provides an additional indication of whether the wind power levels coincide in the two countries. The chromatic spread parameter enables the distribution of wind power levels in the countries to be compared.

The L-based Bland and Altman plot showed that the wind speed was slightly stronger in the United Kingdom. On a monthly basis (over an 8-month observation period), results were all close to zero, indicating that no power transfer was warranted. The L-based Bland and Altman plots for weekly and 3-day analysis showed increased data scatter but fell within the no power exchange boundary (DL) and had a relatively low limit of agreement, supporting the conclusion that power transfer to correct for wind deficiency was not feasible.

The dominant time (H) results showed more than 80% correlation between the UK and Germany wind profiles, with few affected by time shifts between both countries (Sufian et al., 2019), supporting the effective magnitude (L) results (monthly, weekly, 3 days) that power transfer to correct for wind deficiency was not feasible.

The spread results (S) for the United Kingdom and Germany showed that the wind speed was slightly more spread in Germany, but they were correlated to be within the no power exchange boundary (DL) and had a low limit of agreement (Sufian et al. 2019), providing further evidence for the non-feasibility conclusions.

Since the United Kingdom and Germany lie, respectively, at the west and east of a span of 1,500 km across northern Europe, it is reasonable to conclude that the results obtained are relevant to the entire region. The chromatic analysis and Bland and Altman processing indicated a high correlation between wind power-based production in midwinter across northern Europe. Therefore, at that time, when wind is the dominant source of electricity, interconnections will add little to system security.

REFERENCES

Bland, J. M. and Altman, D. (1986). Statistical methods for assessing agreement between two methods of clinical measurements. *Lancet*. 327(8476), 307–310. February 08, 1986.
Fraunhofer Institute. Available at https://www.energy charts. De/power.htm
Renewal Energy Foundation. Available at https://www.ref.org.uk/energy-data
Sufian, A. T., Jones, G.R. and Lawton, J. (2019). *Chromatic Analysis of Wind Generated Power Technical Report*. Center for Intelligent Monitoring Systems. Department of Electrical Engineering an Electronics. The University of Liverpool.

12 Chromatic Comparison of Environmental Factors of High-Voltage Circuit Breaker Gases

G. R. Jones and J. W. Spencer

CONTENTS

12.1 INTRODUCTION

Although SF_6 is a powerful high-voltage insulating gas, it is also recognised as one of the most serious greenhouse warming gases (Stocker et al., 2013; Seeger et al., 2017). Alternatives have been considered from an environmental aspect as well as current interruption capabilities (Kieffel et al., 2016; Seeger et al., 2017). Three gas properties with environmental implications have been suggested as boiling point (BP) (°C), greenhouse warming potential (GWP) and toxicity (TOX) (Uchii et al., 2002; Preve et al., 2015; Gentils et al., 2016; Preve et al., 2016; Kieffel et al., 2016; Seeger et al., 2017). Some preferred pure gases with SF_6 replacement potential are

Fluoroketones ($C_5F_{10}O$)
Fluoronitrile (C_4F_7N)
Carbon Dioxide (CO_2)

Mixtures of some of these gases with air and CO_2 have also been considered.

Values of BP, GWP and TOX have been reported by several investigators for these gases (Seeger et al., 2017).

To illustrate the potential of chromatic processing for comparing the environmental implications of these gases (CO_2, SF_6, $C_5F_{10}O$, C_4F_7N), data for these gases have been processed chromatically and represented on an effective magnitude (L) graph and an X, Y, Z chromatic map.

12.2 DATA NORMALISATION

Numerical values for each of the three environmental parameters (BP, GWP, TOX) reported by investigators (Uchii et al., 2002; Preve et al., 2015; Gentils et al., 2016; Kieffel et al., 2016; Preve et al., 2016; Seeger, 2017) are shown in Table 12.1 for CO_2, SF_6, $C_5F_{10}O$, C_4F_7N.

TABLE 12.1

Values of Normalised BP, GWP and TOX for CO_2, SF_6, $C_5F_{10}O$, C_4F_7N and Nominal Values for Air (Seeger et al., 2017)

GAS	BP (°C)	GWP	TOX (ppmv)
CO_2	−78.5	1	5000
SF_6	−64	23,500	1000
$C_5F_{10}O$	26.5	<1	225
C_4F_7N	−4.7	2100	65
Air	−90 (nominal)	0	0

TABLE 12.2

Values of Normalised BP, GWP and TOX for CO_2, SF_6, $C_5F_{10}O$, C_4F_7N and Nominal Values for Air

GAS	(BP)n	(GWP)n	(TOX)n
CO_2	0.215	0	1
SF_6	0.36	1	0.2
$C_5F_{10}O$	1	0	0.045
C_4F_7N	0.95	0.09	0.01
Air	0.1(nominal)	0	0

These values have been normalised so that they lie within the range 0–1 using the following equations

$$\text{Boiling Point (BP)n} = \left[(100 + BP)/100)\right] \tag{12.1}$$

$$\text{Greenhouse Warming Potential (GWP)n} = (GWP/(GWP)SF_6) \tag{12.2}$$

$$\text{Toxicity (TOX)n} = (TOX/(TOX)CO_2) \tag{12.3}$$

The normalised values for the four pure gases and air using these normalising equations are given in Table 12.2.

12.3   CHROMATIC ANALYSIS

The normalised parameters are treated as chromatic outputs R, G, B (Chapter 1) with chromatic responses R = BP, G = GWP, B = TOX. The R, G, B normalised values have been chromatically transformed (Chapter 1) to produce relative chromatic parameter values for BP, GWP and TOX, which are transformed to X, Y, Z and the effective magnitude L. The results have been compared in two graphs – an effective magnitude (L) versus gas type graph (Figure 12.1) and a relative magnitude X, Y, Z chromatic map (Figure 12.2).

The L versus test gas graph indicates the overall deviation of a gas from the sought-after norm. The X, Y, Z map indicates the relative deviation of each environmental parameter (BP, GWP, TOX) from the ideal and identifying the dominant deviation determining factor.

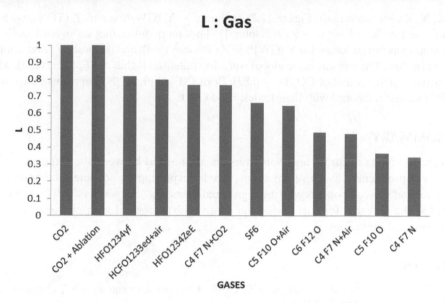

FIGURE 12.1 Effective magnitude of environmental impact (L) versus test gas. ($L = [(BP)n + (GWP)n + (TOX)n]/3$.

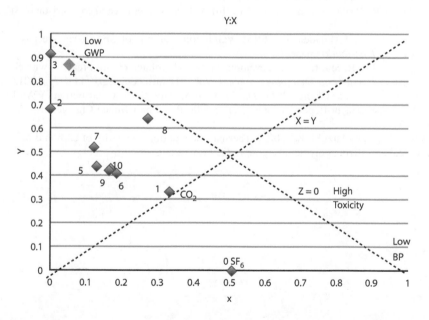

FIGURE 12.2 X: Y: Z chromatic map of the relative magnitude of the three environmental effects for test gases SF_6, CO_2, $C_5F_{10}O$, C_4F_7N. ($X = [100 - (BP)n]/300L$; $Y = [GWP]n/3L$; $Z = [TOX]n/3L$ [$Z =$ Diagonal through origin with $Z = 1$ at $X = Y = 0$]).

12.4   INTERPRETATION

The effective magnitude (L) versus gas type (Figure 12.1) enables the overall deviation of environmental effects of the different gases to be compared. The most well-behaved gas (air) is taken to have the lowest value L value (~0.03), whilst SF_6 is the worst-behaved gas, having a value of 0.52. CO_2 has an L value of 0.405, whilst $C_5F_{10}O$ and C_4F_7N both have lower values of L (0.35) than CO_2 and SF_6 but higher than air.

The X, Y, Z chromatic map (Figure 12.2) is shown as Y (GWP) versus Z (TOX), with X (BP) being indicated as X = 1 − (Y + Z) (Chapter 1). This map shows that an overall well-behaved environmental gas (air) is located at Y (GWP) = Z (Toxicity) $\sim$ 0, and the boiling point temperature (X = 1) is also low. The worst relative global warming potential is that of SF_6 (Y = 0.64), whilst the worst relative toxicity is that of CO_2 (Z = 0.82). Both $C_5F_{10}O$ and C_4F_7N appear to have relatively high X(BP) values compared with their toxicity and GWP.

12.5 SUMMARY

The chromatic results illustrate how compromises are needed between the three environmental factors. As a consequence, investigations are in place for exploring the possible compensating effects of mixing various gas components with these gas candidates (e.g., Seeger et al., 2017). The chromatic approach has the potential for quantifying the relative advantages of such mixtures in addressing the environmental limitations.

REFERENCES

Gentils, F., Maladen, R., Piccoz, D., and Preve, C. (2016). Load break switching in SF_6 alternative gases for MV applications, Schneider Electric

Kieffel, Y., Irwin, T., Ponchon, P., and Owens, J. (2016) Green gas to replace SF_6 in electrical grids. *IEEE Power and Energy Magazine*, Vol 14, Issue 2, pp. 32–39

Preve, C., Maladen, R., Piccoz, D., and Biasse, J. (2016). Validation method and comparison of SF_6 alternative gases, CIGRE

Preve, C., Piccoz, D., and Maladen, R. (2015). Validation methods of SF_6 alternative gas 23 Int. Conf. *On Electricity Distribution* paper 0493

Seeger, M. et al. (2017). Recent trends in development of high voltage circuit breakers with SF_6 alternative gases. *Plasma Physics and Technology Journal*, Vol. 4(1), http://doi.org/10.14311/ppt.2017.1.8

Stocker et al. (2013). *Climate Change 2013: The Physical Science Basis*. Contribution of Working Group 1 to the Fifth Assessment Report of Intergovernmental Panel on Climate Change, https://www.ipcc.ch/report/ar5/wg1/.

Uchii, T., Shinkai, T., and Suzuki, K. (2002). Thermal interruption capability of carbon dioxide in a puffer circuit breaker utilizing polymer ablation, *IEEE PES T&D Conf*

13 Chromatic Line of Sight Particle Monitoring

A. T. Sufian and J. W. Spencer

CONTENTS

13.1 INTRODUCTION

There has been increasing concern over airborne microparticles in the environment and the potential health risks they produce (Nobel and Prather, 1998; Holgate et al., 1999). Detection of such particles in real-world conditions has traditionally been done with laser light scattering using optical particle counters (OPCs) (Welker, 2012). However, the use of free space laser beams can be restrictive with regard to the location of such systems. A technique which avoids the need for using laser light has been successfully used and was based on the use of a polychromatic light source to address particles captured on part of a particle filter which was chromatically analysed. Part of the filter was used as a reference area (Reichelt et al., 2006; Jones et al., 2008). The system was further developed to monitor airborne particle pollution using urban closed-circuit television camera (CCTV) networks for monitoring the particle capturing unit remotely (Kolupa et al., 2010). The chromatic technique has more recently been used as a line-of-sight approach based on combining a narrow-band chromatic source with a polychromatic source and an optical path several meters long in free air (Sufian and Spencer, 2018).

13.2 CHROMATIC MONITORING SYSTEM

The monitoring system consisted of a white-light light emitting diode (LED) source, chromatic filters positioned in front of it and a red LED source placed near a camera emitting across the optical path between the camera and white LED source (Figure 13.1) (Sufian and Spencer, 2018). Particles near the camera passing through the red LED beam side scatter light into the camera. Particles in the optical path between the white light source and the camera backscatter the light. The chromaticity of the source could be varied by chromatic filters placed in front of the source for changing the sensitivity of the camera to the light affected by the particles.

Optical filters in the form of white paper placed in front of the source were used to diffuse the light intensity, along with various forms of transparent chromatic filters. Furthermore, a semicircular black paper shutter carrying various aperture geometries was used in front of the camera to govern the shape of the light image captured by the camera (Figure 13.2). A semicircular arrangement of a black paper filter covering the lower part of the source (Figure 13.2a) was used. Alternatively, the same arrangement with an orange translucent plastic filter covering the upper and lower parts of the optical source was used (Figure 13.2b). A rectangular aperture produced by a combination of two black filters covering the upper and lower parts of the source and an orange translucent filter in

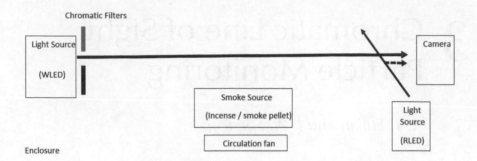

FIGURE 13.1   Schematic diagram of the particle monitoring system (Sufian and Spencer, 2018).

FIGURE 13.2   Example of images obtained with different arrangement of chromatic filters (Sufian and Spencer, 2018). (a) Images with white and black paper filters to form a semicircular aperture. (b) Images with white, orange and black paper filters to form a semicircular aperture. (c) Images with white, black and orange filters arranged to form a rectangular aperture. (d) Images with white, black and neutral-density filters arranged to form a rectangular aperture superimposed upon a semicircular aperture.

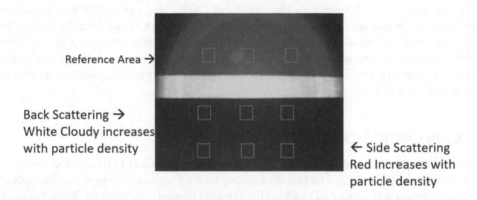

FIGURE 13.3   Different monitoring locations (Sufian and Spencer, 2018).

between has been used (Figure 13.2c), as well as a combination of a black filter covering the lower part of the source and a neutral-density filter over the upper part with white light transmitted in between (Figure 13.2d).

Smoke particles produced by incense sticks (particle size 1–2.5 microns) affected the images. As the particle density increased, cloudy white and red areas in parts of the image also increased. When the images were chromatically analysed, each showed a change in chromaticity, with the arrangement of Figure 13.2d showing the greatest change. Further tests were performed with various monitoring points on the image (Figure 13.3) addressing the side-scatter of the red LED beam from particles and the backscatter from the white-light LED. As the particle concentration increased, the back- and side-scattered light also changed.

13.3 CHROMATIC TEST RESULTS

Changes in the optical signals produced by different levels of particles were analysed chromatically from the camera R, G, B outputs at various locations on the images (Figure 13.3). Chromatic parameters L (lightness) and S (saturation) (Chapter 1) were calculated for the various particle conditions. Further chromatic parameters (Lr and L′, S′) were also derived from the information in the more restricted long- and medium–short-wavelength range of the visible spectrum, respectively, where Lr = R (intensity strength of the side-scattered red light) and L′ = G + B/2, S′ = (G − B)/(G + B) (intensity and spread of the backscattered midrange light).

The variation of the chromatic parameters L and S with airborne particle concentrations is shown in Figure 13.4. Figure 13.4a shows the variation of L and S for the reference area (Figure 13.3) to be independent of the particle concentration. Figure 13.4b corresponding to the side-scattered red light (Figure 13.3) shows that both the red channel intensity (Lr) and saturation (S) increase monotonically with particle concentration but with S changing more than Lr. Figure 13.4c (corresponding to the backscattered light) shows that both the lightness (L′) and saturation (S′) increase monotonically with particle concentration, albeit at a relatively low level and similarly to each other.

The implication of these results is that the saturation of the side-scattered light (S′) provides a reliable parameter for tracking airborne microparticles without the need to use a particle filter to capture the particles. The constancy of the reference area under controlled experimental conditions also confirms the reliability of the approach.

Further chromatic analysis of the captured data is also possible. For example, chromatic information (L′, S′) may be extracted from the medium–short-wavelength range of the backscattered light and normalised for comparison. The result of such a procedure is shown in Figure 13.5 for

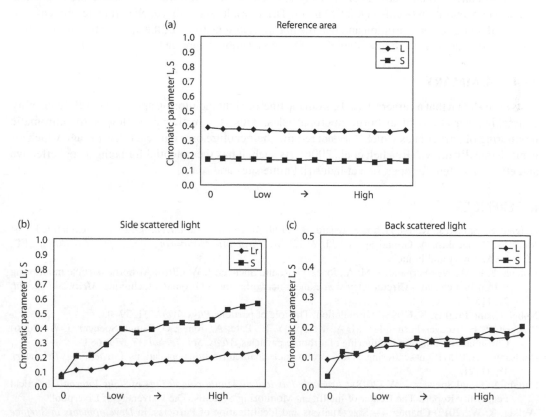

FIGURE 13.4 Calibration graphs for various chromatic parameters (Sufian and Spencer, 2018). (a) Reference area; (b) side-scattered light; (c) backscattered light (Sufian and Spencer, 2018).

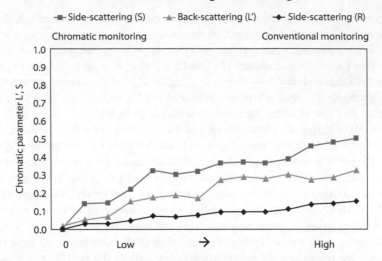

FIGURE 13.5 Comparison of normalised trends for different parameters (side-scattering chromatic [S], backscattering chromatic [L′], side-scattering [R output]) (Sufian and Spencer, 2018).

medium–short-wavelength backscattered light (L′), the saturation (S) and the value of the intensity of the long-wavelength output (red-channel Lr as a function of particle concentration). The latter represents conventional methods of optically monitoring monochromic light scattering of particle and dust concentration (Renlsang, 2015) in air. These results show that each parameter increases in value with particle concentration and that the normalised sensitivity of the side-scattered saturation (S) and backscatter (L′) is higher than that of the conventional approach.

13.4   SUMMARY

Tests have shown that a camera-based chromatic line-of-sight monitoring approach has the capability of tracking variations of airborne microparticles. This is a further evolution of the chromatic monitoring of particulates which does not require their capture on a particle filter through which air is mechanically drawn (Reichelt et al., 2006). As such, it has the potential for being a cost-effective and efficient system for operation at industrial mine sites and so on.

REFERENCES

Holgate, S., Samet, J., Koren, H. and Maynard, R. (1999) *Air Pollution and Health*. London: Academic Press.
Jones, G. R., Deakin, A. G. and Spencer, J. W. (2008) *Chromatic Monitoring of Complex Conditions*. FL, USA: Taylor and Francis.
Kolupa, Y. E., Aceves-Fernandez, M. A., Jones, G. R. and Spencer, J. W. (2010) Airborne particle monitoring with urban closed – Circuit television camera networks and a chromatic technique. *Meas. Sci Techol.* 21, 115204.
Nobel, C. and Prather, K. (1998) Air pollution: The role of particle. *Phys. World* 11, 39–43.
Reichelt, T. E., Aceves-Fernandez, M. A., Kolupa, Y. E., Pate, A., Jones, G. R. and Spencer, J. W. (2006) Chromatic modulation monitoring of airborne particles. *Meas. Sci. Techol.* 17, 675–683.
Renlsang, X. (2015) Light scattering: A review of particle characterization applications. *Particuology (Elsevier)*, 18, 11–21.
Sufian, A. T. and Spencer, J. W. (2018) *Chromatic Particulate Monitoring at Line of Sight*. Internal Technical Feasibility Report. The Center of Intelligent Monitoring Systems, The University of Liverpool.
Welker, R. W. (2012) Chapter 4—Size Analysis and Identification of Particles, in *Developments in Surface Contamination and Cleaning*, R. Kohli and K. L. Mittal, Eds. Oxford: William Andrew Publishing, pp. 179–213.

Section V

Advanced Chromatic Monitoring

This section describes more advanced applications of the chromatic monitoring approach. It considers adaptation of chromatic methods for addressing multidimensional signals (three-dimensional space monitoring), the use of primary and secondary chromatic analysis for monitoring partial discharges and switchgear contact wear, acoustic frequency monitoring of high-voltage transformers and three-phase power transmission lines. It also addresses problems of combining sets of data from a range of conventional measurements on high-voltage transformer oils and assessing the performance of various gases used in high-voltage circuit breakers.

Section V

Advanced Chromatic Monitoring

14 Introduction to Advanced Chromatic Analysis

A. A. Al-Temeemy, J. W. Spencer and G. R. Jones

CONTENTS

14.1 INTRODUCTION

Experience with the chromatic approach has shown that it provides a means for processing a signal that avoids some of the difficulties associated with other methods, including the complexity of transformed Fourier signals and transformation instability in the presence of noise (Jones et al., 1996; Al-Temeemy and Spencer, 2010, 2015a,b). The generic nature of the chromatic approach enables it to be extended to many domains of information extraction (Al-Temeemy and Spencer, 2014, 2015a,b) (e.g., continuous signals, optical signals etc.). It can overcome difficulties of processing methods which use a surfeit of information with excessive computational difficulties of processing times and costs.

More advanced forms of chromatic monitoring and analysis have evolved which include the application of the R, G, B chromatic processors for addressing a particular signal in multiple signal domains and additionally with R, G, B processors having different responses.

14.2 PROPERTIES AND ADAPTABILITY OF CHROMATIC PROCESSORS

14.2.1 CHROMATIC PROCESSOR PROPERTIES

Two examples of the deployment of three chromatic processors adjusted for addressing a well-defined signal are shown in Figure 14.1a(i) and b(i). In both cases, the three processors are adjusted to cover only the extent of the signal. Figure 14.1a(i) shows the use of three Gaussian profile sensors $R(lo)$, $G(lo)$, $B(lo)$ with the extreme sensor $R(lo)$ divided to cover each end of the signal distributed along a location lo. Figure 14.1b(i) shows the same signal addressed by three triangular processors (R, G, B) with $R(lo)$ and $B(lo)$ as the extreme sensors and the middle sensor $G(lo)$ covering the whole signal width but with half the amplitude of $R(lo)$, $B(lo)$.

As a result of these features, the combination of the two sets of R, G, B processors profiles differs from those already discussed in Chapter 1 (all equal amplitude; no single processor covering each end of the signal). The dominant value (H) of a monochromatic signal (l) tracked across each of the two sets of R, G, B processors is shown in Figure 14.1a(ii) and b(ii). This shows that both sets of processors have an almost monotonic variation of H with l but with the Gaussian profiles having

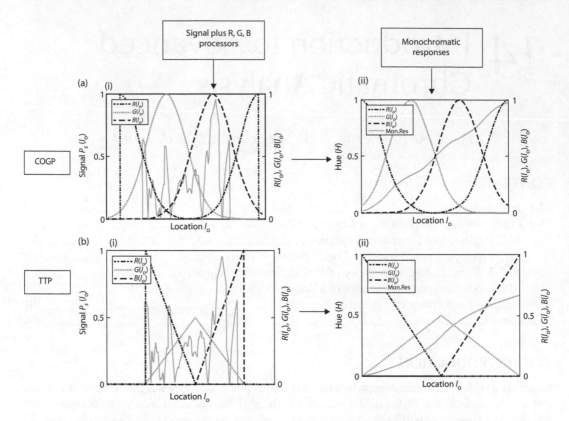

FIGURE 14.1 Chromatic processors and their monochromatic responses. (a(i)) COGP = Continuous overlapping Gaussian processors. (b(i)) TTP = Truncated triangular processors. (a(ii)) Tracked monochromatic signal monitored by COGP. (b(ii)) Tracked monochromatic signal monitored by TTP.

a higher sensitivity ($0 \rightarrow 1$ compared with 0.66). These variations may be compared with those for the original processors described in Chapter 1. The Gaussian-type processor is referred to as a continuous overlapping Gaussian processor (COGP), whilst the triangular processor is referred to as a truncated triangular processor (TTP).

14.2.2 Adaptation of Processor Locations and Widths

For many applications, it is necessary to match the R, G, B processors' widths and locations to a signal's width and location. Figure 14.2 shows a signal $P_R(lo)$ distributed along a location lo, along with three non-orthogonal processor responses $R(lo)$, $G(lo)$, $B(lo)$.

The centroid (Cm) of the signal $P_R(lo)$ of length ℓ is determined from Al-Temeemy and Spencer (2015a,b).

$$Cm = Sum\ of\ (lo\ P_R(lo))/Sum\ of\ (P_R(lo))$$

This is then used to determine the centres and widths of each of the three chromatic processors (R, G, B).

14.2.3 Normalised Chromaticity of Two-Dimensional Signals

A two-dimensional signal (e.g., optical image) may be addressed by adapting the chromatic processors' locations and width, as indicated in Section 14.2.2, but in two dimensions. Displaced and magnified images may then be compared (Figure 14.3) to check for identification

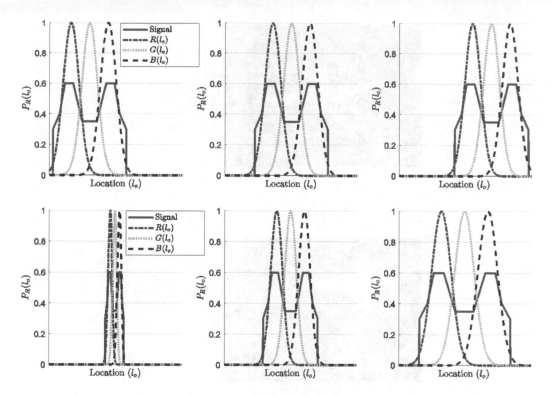

FIGURE 14.2 Adaptations of the locations and widths for Gaussian processors with the input signal.

This is achieved by first adjusting the locations and widths of the chromatic processors (R, G, B) in the x dimension to calculate the chromatic parameters $H(x)$ and $S(x)$. The process is then repeated for the y dimension to yield $H(y)$ and $S(y)$. The image is then characterised by the four parameters $H(x)$, $S(x)$, $H(y)$, $S(y)$. For the example shown in Figure 14.3, $H(x)$ was consistent to within 1.2%, $S(x)$ to within 1.0%, $H(y)$ to within 2% and $S(y)$ to 0.4%. These results indicate a high level of invariance for such distortions.

14.2.4 Normalised Chromaticity of a Rotated Signal

Rotational displacements of a two-dimensional signal can also be accommodated via two-dimensional chromaticity. The chromatic processors in this case are used after the radon transformation of the image $f(xn, yn)$ at a normalising angle (Al-Temeemy and Spencer 2015a,b) (Figure 14.4).

For the example shown in Figure 14.4, such a process leads to values of chromatic parameters compared to those of the original image to within 0.8% for $H(x)$, 0.1% for $S(x)$, 2% for $H(y)$ and 0.1% for $S(y)$.

14.3 SUMMARY AND OVERVIEW

Extensions of the basic chromatic approach which have already been exploited for a variety of applications have been described and have the potential for further exploitation. One aspect takes advantage of a variety of different forms of the three basic non-orthogonal processors. This enables forms of the processor responses to be adjusted for enhancing performance for particular applications.

Also, chromatic processing has been expanded into a number of signal dimensions, which enables the combination of chromatic parameters from several domains to be formalised. At its simplest level, primary chromatic parameters may be further processed as a function of time and so on, providing a wider choice of chromatic parameters for extracting the required information.

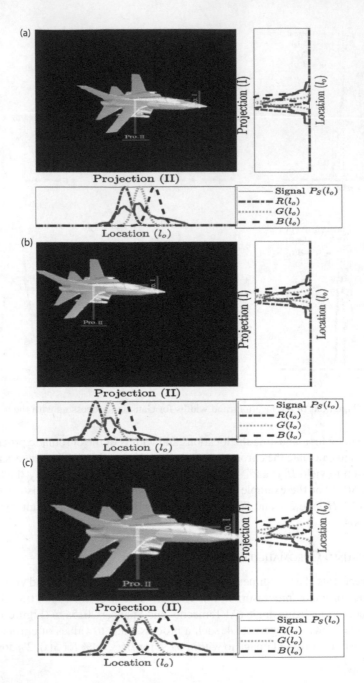

FIGURE 14.3 Normalised projections of fighter jet images. (a) Original image. (b) Shifted to left. (c) Enlarged one and a half.

This has enabled time and frequency domain chromatic processing to be achieved. It has also enabled the chromatic processing of two length dimension (x, y) distributed signals to be addressed to quantify the shape of an image.

A primary chromatic analysis involves addressing one signal dimension (e.g., frequency distribution) with three chromatic processors (R, G, B) to produce primary chromatic parameters x, y, z, L, H, S (Chapter 1). The variation of these parameters within a second signal dimension (e.g., time) may then be addressed with secondary chromatic processors $R(2), G(2), B(2)$ to yield secondary chromatic

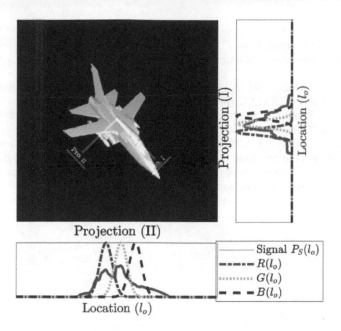

FIGURE 14.4 Radon transformation for rotated image.

parameters. This provides a choice of several chromatic parameters for quantifying a signal. Examples of such a procedure and their application include the time–frequency analysis of partial discharges (Chapter 18) and the time–wavelength analysis of electric arc contact wear in a high-voltage current interrupter (Chapter 17). A similar approach may be deployed to address a signal in three spatial dimensions whereby primary chromatic analysis is performed individually for each dimension and secondary chromatic parameters produced from a combination of the primary chromatic parameters (Chapter 15). A further development has been the analysis of three interrelated parameters (e.g., fault detection in a three-phase electric power transmission system; Chapter 19). There is also the deployment of three primary chromatic processors (R, G, B) for addressing the simultaneous magnitudes of a large number of different parameters organised into three representative groups (e.g., dissolved gases and so on in transformer oil in Chapter 20 and the performance of different gases in high-voltage circuit breakers in Chapter 21). A chart highlighting these adaptations of chromatic analysis is given in Figure 14.5.

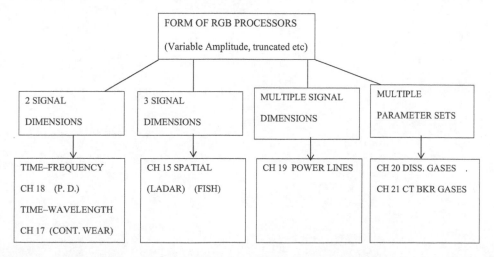

FIGURE 14.5 Fundamental basis of various forms of advanced chromatic analyses.

REFERENCES

Al-Temeemy, A. A. and Spencer, J. W. (2010). Invariant spatial chromatic processors for region image description. *IEEE International Conference on Imaging Systems and Techniques*, pp. 421–425, 10.1109/IST.2010.5548524

Al-Temeemy, A. A. and Spencer, J. W. (2014). Laser radar invariant spatial chromatic image descriptor. *Opt. Eng.*, 53(12), 10.1117/1.OE.53.12.123109

Al-Temeemy, A. A. and Spencer, J. W. (2015a). Invariant chromatic descriptor for LADAR data processing. *Mach. Vis. Appl.*, 26(5), pp. 649–660, 10.1007/s00138-015-0675-0

Al-Temeemy, A. A. and Spencer, J. W. (2015b). Chromatic methodology for laser detection and ranging (LADAR) image description. *Optik,* 126(23), pp. 3894–3900, 10.1016/j.ijleo.2015.07.182

Jones, G. R. et al. (1996). Chromatic processing of optoacoustic signals for identifying incipient faults on electric power equipment. *IEE Colloquium on Intelligent Sensors (Digest No: 1996/261)*, Leicester, UK, pp. 3/1–3/5.

15 Chromatic Monitoring of Spatial Dimensions

A. A. Al-Temeemy, J. W. Spencer and L. U. Sneddon

CONTENTS

15.1 INTRODUCTION

Enhanced chromatic monitoring procedures have been deployed with various types of signal production hardware (visible optical, infrared [IR] detection, lasers etc.), various signal processing methods (ViBe, radon etc.) and different physical dimensions (one–three dimensions). Three-dimensional chromatic processing for laser detection and ranging (LADAR) has been developed. A combination of two-dimensional polychromatic light imaging with three passive infrared sensors has been produced for cost-effective, non-intrusive room monitoring. Three-dimensional spatial monitoring has been produced for addressing fish clustering.

15.2 LASER DETECTION AND RANGING SYSTEM MONITORING

Radio detection and ranging (RADAR) is the process of transmitting, receiving, detecting and processing an electromagnetic wave reflected from a target. RADAR techniques have expanded into many modern applications areas and have also moved into the optical portion of the electromagnetic spectrum. Using lasers as optical transmission sources, a specific category of optical RADAR systems has been proposed called laser radar or laser detection and ranging (LADAR) (Al-Temeemy and Spencer, 2014). LADAR are three-dimensional (3D) spatial measurement systems. The power of these systems lies in the inherent 3D nature of the data that are produced (Al-Temeemy and Spencer, 2015a). They are an attractive alternative to RADAR systems, because through the use of optical laser wavelengths (which are shorter than RADAR wavelengths), they provide very high-resolution three-dimensional images. In addition, light velocity allows LADAR systems to take numerous measurements per second (Al-Temeemy, 2017). LADAR images are created by scanning a scene with laser beams; the scanning angles and the return time for these beams are used to calculate a 3D-range image, which in turn represents the spatial location of the intersection of the laser beam with the scanned scene (Al-Temeemy, 2017). LADAR systems have diverse applications (Al-Temeemy and Spencer, 2015a), including quality control, surveying, mapping, terrain characterisation, safety monitoring, disaster reconnaissance etc. (Al-Temeemy and Spencer, 2015b). A LADAR prototype system with its laser ranging and scanning unit is shown in Figure 15.1. (Al-Temeemy and Spencer, 2011).

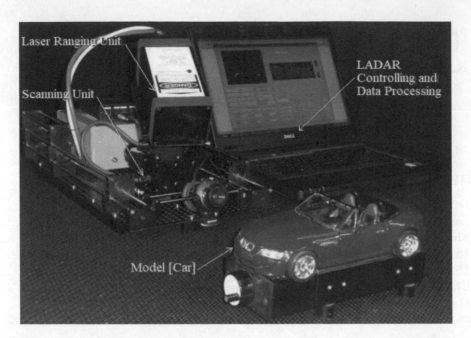

FIGURE 15.1   LADAR prototype scanning model car.

Different methods were developed for processing the incoming information from LADAR monitoring systems. Some methods describe LADAR images based on shape histogram methods which describe LADAR images as histograms of point fractions. These methods generally suffer from the presence of noise and require a high number of features. This increases the recognition time and requires more memory to store the features (Al-Temeemy and Spencer, 2014).

Others use robust descriptors for LADAR images such as moment invariants. Their general disadvantage is noise sensitivity, which limits and restricts their applications (Al-Temeemy and Spencer, 2015a). Furthermore, surface properties have been used with LADAR images such as normals and regional shape. Surface normals are sensitive to noise, while the regional shape extraction approach is robust to noise, but it is computationally expensive, requires a large amount of space to store the features and has low discriminating capability (Al-Temeemy and Spencer, 2015b).

Advanced chromatic methods have been used because of their simplicity, high discrimination capability, noise insensitivity and affine (rotation, translation and scaling) invariant capability. These chromatic methods are based on extracting a robust feature that is relatively unaffected by the noise which usually disturbs LADAR measurements. This feature is the silhouette image of the 3D LADAR data from its perspective view (Al-Temeemy and Spencer, 2014, 2015a,b). As an example, Figure 15.2 shows the image addressing features. Figure 15.2a shows an image of a fighter jet being addressed by a LADAR system. Figure 15.2b shows the point cloud of this fighter from a rotated view. Figure 15.2c shows it from the LADAR perspective. Figure 15.2d shows the resultant silhouette of the fighter. While the range data (Figure 15.2b) shows significant noise, the LADAR view and its resultant silhouette shown in Figure 15.2d show a relatively smooth image. This smoothness comes from the high pointing accuracy of the scanning unit in comparison to the distortion of the range measurement (Al-Temeemy and Spencer, 2015b).

The resultant silhouette image is then processed by extracting the normalised projections (which are rotation invariant) using radon transform and then processing these projections with a special type of chromatic processors called shift- and scale-invariant spatial chromatic processors. The processors' deployments in this type adapt their locations and widths with respect to these projections, which produces chromatic values invariant to rotation, translation and scale effects (Al-Temeemy and Spencer, 2010, 2014, 2015a,b) as shown in Figures 14.3 and 14.4 (Chapter 14).

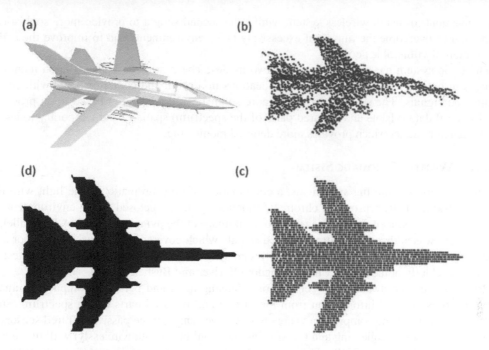

FIGURE 15.2 Silhouette image generation from distorted LADAR data. (a) Fighter jet model, (b) scan data from rotated view, (c) perspective view, (d) resultant silhouette image.

Methods for normalising images of a displaced, magnified and rotated object have been described in Chapter 14 (Figures 14.3, 14.4). These advanced chromatic methods have been evaluated with simulated LADAR data using special software called a LADAR simulator (Al-Temeemy, 2017). This simulator models each stage from the laser source to the data generation and so is able to generate simulated LADAR data through scanning 3D computer aided design (CAD) models with different artefacts such as noise, resolution, view, scaling, rotation and translation (Al-Temeemy and Spencer, 2015b).

The simulation results show high discrimination capability for the advanced chromatic methods over the moments invariant, which have also been used to benchmark the results. The results also show constant performance during scaling, rotation and noise effects, which shows the effectiveness of these methods and their robustness with noise effects (Al-Temeemy and Spencer, 2015b). Advanced chromatic methods were also evaluated with real LADAR data. The experimental results show similar general behaviour compared to the simulated data, which proves the ability of the approach to process real LADAR data and provide recognition rates higher than traditional techniques like the invariant moment descriptor (Al-Temeemy and Spencer, 2015b).

These advanced chromatic methods are simple, and their discrimination capability can be easily extended by either increasing the number of spatial chromatic processors or using additional normalised projections. The small number of chromatic features for these methods means that they require less storage space and processing time (Al-Temeemy and Spencer, 2010, 2014, 2015a,b).

15.3 MULTISPECTRAL DOMICILIARY HEALTHCARE MONITORING

There is an increasing need for independent living at home by the elderly population (Al-Temeemy, 2018, 2019). To alleviate operational and financial difficulties which result from this and to provide comfortable living conditions, new monitoring systems are needed. Several methods have been proposed. One approach has been with a system using a combination of visible and infrared light for chromatically addressing a living accommodation (Jones et al., 2008; Al-Temeemy, 2018, 2019). The approach is versatile and has evolved in two major ways. The first was to produce a cost-effective and convenient-to-install wireless system, whilst the second sought to provide more sophisticated adaptation to overcome the impact of excessively noisy environments and to improve the activity recognition of vulnerable people.

The basic system can be used in one of two modes. The first mode is based upon monitoring changes in optical chromaticity of various locations in a room and combining this with changes in infrared signals. The second mode is a more sophisticated adaptation based upon processing multispectral data (visible and infrared parts of the spectrum) spatially and temporary to identify the human silhouette, which provides more detailed monitoring.

15.3.1 Wireless Chromatic System

The basis of the monitoring system was a combination of polychromatic visible light with three passive infrared (PIR) sensors for chromatic monitoring in an enclosed living environment. The optical monitor produced a two-dimensional optical image of the living environment. In parallel, this environment was addressed by three infrared signals which covered three overlapping floor areas, producing seven distinguishable areas. Such systems have been successfully installed and tested at care for the elderly homes in the United Kingdom (Driver and Busfield, 2005).

Recent system hardware setup consists of a sensing head and personal computer to analyse incoming sensing head information from the visible and infrared parts of the spectrum (Smith, 2019). The sensing head comprises a visible band image sensor, three passive infrared sensors and a microcontroller (for collecting and transferring the monitoring data wirelessly) with the required electronics (Al-Temeemy, 2019). The sensing head for the monitoring system with its internal structure is shown in Figure 15.3.

Recent developments of the system involved the optical output signals being processed locally with optical chromaticity, whilst the PIR output signals were also processed locally but via spatial chromaticity. The processing was undertaken within the small monitoring unit shown in Figure 15.3. The optical and spatial chromaticities were then processed in combination to provide information not only about movement but also particular location conditions (e.g., closed/opened doors, television on/off etc.) and human activities. The values of the basic chromatic parameters were transferred wirelessly to a central control hub, the procedure having reduced the amount of data needing to be transferred compared with the previous hardwired system (Jones et al., 2008; Smith, 2019). As a result of the chromatic data compression, a central processing unit was developed which was capable of supporting 16 sensor head units at different ceiling locations in different rooms, all operating in parallel and continuously 24/7.

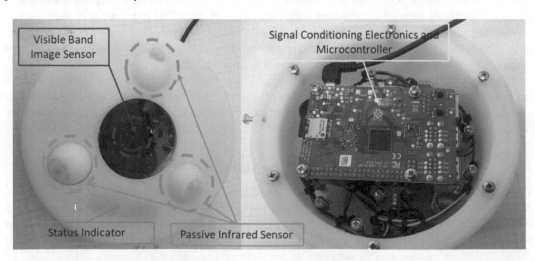

FIGURE 15.3 Sensing head and internal structure of the monitoring system.

15.3.2 ADVANCED CHROMATIC INFORMATION PROCESSING

The advanced monitoring of vulnerable people involves using a chromatic monitoring system (Section 5.2.3.1) to acquire optical and infrared signals, followed by the use of two new monitoring stages to recognise people's activity. The first stage is for identifying the silhouette image of person to be monitored, while the second stage processes this image to recognise the living activity. Extracting and identifying the silhouette image is achieved by using a foreground detector such as ViBe (Al-Temeemy, 2018). Identifying silhouettes in noisy environments with illumination changes and dynamic backgrounds is difficult (Al-Temeemy, 2019). However, with the present system, both sensor types (visible and IR) will respond when the human moves across a specific location such as that shown in Figure 15.4 (top). Thus, the developed system generates the spatial-temporal probability of detection from the PIR sensors and then correlates this with the foreground output generated from the visible sensor output (Figure 15.4 [top]) using the ViBe detector. The result of the correlation is then

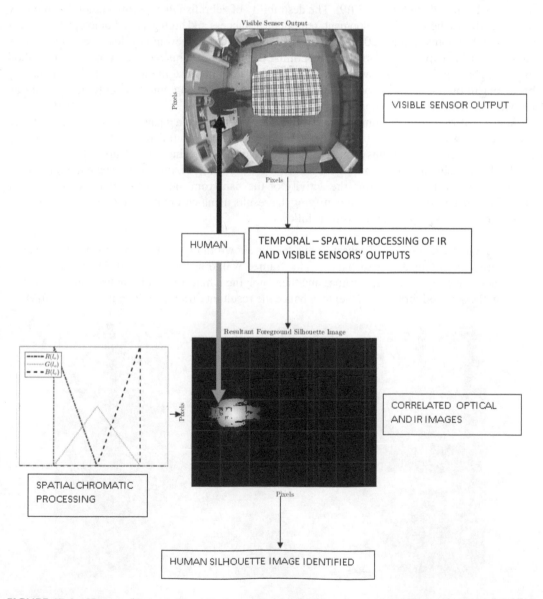

FIGURE 15.4   Human silhouette identification with a combination of optical and IR sensing.

processed by spatial chromatic processors (Figure 15.4 [bottom left]) to locate the silhouette region that corresponds to the monitored person and extract it from the other regions (Figure 15.4 [bottom]).

The identified silhouette image is then processed by different types of invariant spatial chromatic processors to recognise the living activity. Experimental data sets have been used to evaluate the performance of the chromatic processing and silhouette identification methods. The results show that the use of chromatic methodology can efficiently deal with events that disturb the monitoring systems. They also show better performance in comparison to traditional methods when describing daily living activity (Al-Temeemy, 2018, 2019).

15.4   CHROMATIC MONITORING OF GROUPS OF ZEBRAFISH (*DANIO RERIO*)

Fish are finding increased use as model species within a variety of biomedical and neurobiological contexts. Zebrafish (*Danio rerio*) are estimated to account for 50% of the total number of fish used (UK Home Office Report, 2019). The desirability of zebrafish in experimentation has recently increased due to their rapid development, reproductive success and high genetic homology to humans (80%–85%) (Thomson et al., 2019). Researchers are required to prevent any negative states such as pain when using experimental animals to optimise their welfare. This requires a convenient method to detect abnormal behaviour, which is difficult to achieve because the identification of abnormal behaviour in one or a few fish within a larger group is challenging, and thus only information on individual zebrafish exists (Thomson et al., 2019).

Analysis of video frames from an electronic camera has provided a means for developing intelligent chromatic software (chromatic fish analyser; CFA), to monitor the overall average behaviour in a group of zebrafish, some with their fins clipped and subjected to pain-relieving drugs. Chromatic fish analysis involved addressing a tank containing the fish with a video camera inclined, as shown in Figure 15.5.

The CFA involves calculating the activity of the fish from the video frames, chromatically analysing the activity images and quantifying the results using chromatic maps (Figure 15.6).

The three analysis procedures were as follows:

- A typical video image of zebrafish in a tank is shown in Figure 15.6. An image of the fish activity is obtained from such input video frames by finding the absolute difference between successive recorded video frames and then applying a hard-thresholding technique (based on the selected threshold value) to enhance the resultant difference (Thomson et al., 2019).

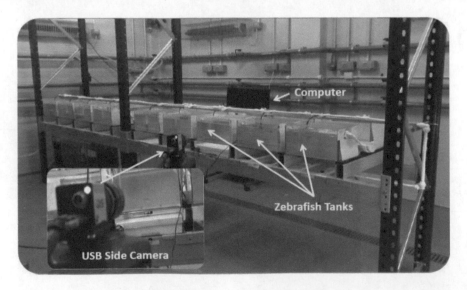

FIGURE 15.5   Chromatic monitoring of zebrafish tanks using USB video camera.

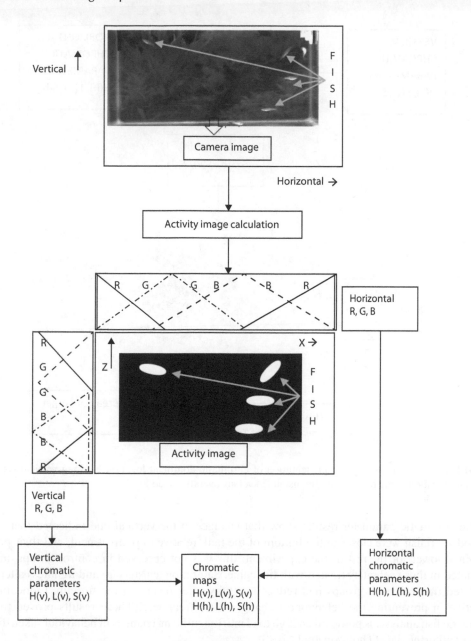

FIGURE 15.6 Block diagram for chromatic monitoring of zebrafish.

- Chromatic addressing involves applying two sets of spatial chromatic processors on each resultant activity image (one vertically and the other horizontally deployed; Figure 15.6). The spatial response of these processors is chosen to provide uniform sensitivity across the entire video frame (Al-Temeemy, 2018, 2019).
- The processors' outputs for each image are transformed into horizontal and vertical chromatic parameters of hue, saturation and lightness which reflect the behaviour of the group of zebrafish for that image. Hue indicates the dominant location and height of the group, saturation the spread of the group and lightness the activity level. (Thomson et al., 2019).

The 3D representations of the chromatic parameters for each image frame are shown in Figure 15.7 for H(vertical): H(horizontal): S as clouds of points.

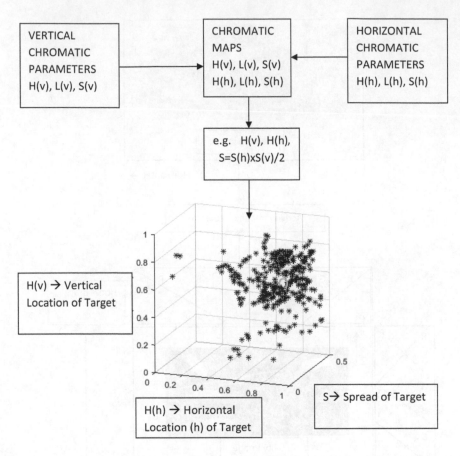

FIGURE 15.7 Example of 3D representation of chromatic dominant location H along vertical (v) versus dominant location along horizontal dimension (h) versus overall spread S.

The chromatic parameter results show that changes in the vertical hue indicated that all fin-clipped zebrafish were closer to the bottom of the tank relative to pretreatment, and their position remained lower for the rest of the experiment; this was not observed in control groups and was alleviated in those zebrafish treated with lidocaine. Saturation (clustering) and lightness alterations indicated that zebrafish groups had reduced activity after receiving the fin clip. Lidocaine was effective in preventing the behavioural changes after treatment. These results proved that the chromatic fish analyser is powerful enough to identify significant changes in behaviour taken directly from monitoring data (Thomson et al., 2019).

15.5   OVERVIEW AND SUMMARY

Examples of the deployment of multidimensional monitoring in three-dimensional space have been considered. Three different applications are described: LADAR-based monitoring, care of the elderly monitoring and fish tank monitoring. In each case, the chromatic approach has provided an efficient, cost-effective and convenient-to-use approach with good levels of performance compared with other monitoring methods. The method is flexible and has potential for extrapolation to other space domain monitoring applications.

REFERENCES

Al-Temeemy, A. A. (2017) The development of a 3D LADAR simulator based on a fast target impulse response generation approach. *3D Research*, Springer, 8, 31. ISSN 2092-6731, doi: 10.1007/s13319-017-0142-y.

Al-Temeemy, A. A. (2018) Human region segmentation and description methods for domiciliary healthcare monitoring using chromatic methodology. *J Electron Imaging*, SPIE 27, 27–14, doi: 10.1117/1. JEI.27.2.023005.

Al-Temeemy, A. A. (2019) Multispectral imaging: Monitoring vulnerable people. *Optik – International Journal for Light and Electron Optics*, Elsevier, 180, 469–483, doi: 10.1016/j.ijleo.2018.11.042.

Al-Temeemy, A. A. and Spencer, J. W. (2010) Invariant Spatial Chromatic Processors for Region Image Description. *EEE International Conference on Imaging Systems and Techniques (IST)*, Thessaloniki, Greece, 421–425.

Al-Temeemy, A. A. and Spencer, J. W. (2011) Three-Dimensional LADAR Imaging System using AR-4000LV Laser Range-Finder. *Proceeding of SPIE – Optical Design and Engineering IV, Marseille*. France, (8167), 816721-(1-10).

Al-Temeemy, A. A. and Spencer, J. W. (2014) Laser radar invariant spatial chromatic image descriptor. *SPIE-Opt. Eng.* 53(12), 123109.

Al-Temeemy, A. A. and Spencer, J. W. (2015a) Invariant chromatic descriptor for LADAR data processing. *Machine Vision and Applications*, 26(5). Springer, doi: 10.1007/s00138-015-0675-0.

Al-Temeemy, A. A. and Spencer, J. W. (2015b) Chromatic methodology for laser detection and ranging (LADAR) image description. *Optik – International Journal for Light and Electron Optics*, Elsevier, 126(23), 3894–3900.

Driver, S. and Busfield, R. (2005) *Evaluation Project Report; Merton Intelligent Monitoring System (MIMS)*. Evaluation Project Report, Roehampton University, UK.

Home Office Report (2019). *Annual Statistics of Scientific Procedures on Living Animals, Great Britain 2018*. ISBN 978-1-5286-1336-1.

Jones, G. R., Deakin, A.G., and Spencer, J. W. (2008) *Chromatic Monitoring of Complex Conditions*. CRC Press, Taylor and Francis Group, ISBN 13:978-1-58488-988-5

Smith, D. H. (2019) CIMS Internal, Report.

Thomson, J. S., Al-Temeemy, A. A., Isted, H., Spencer, J. W., and Sneddon, L. U. (2019) Assessment of behaviour in groups of zebrafish (*Danio rerio*) using an intelligent software monitoring tool, the chromatic fish analyser. *J Neurosci Methods*, Elsevier, 328, 108433, doi: 10.1016/j.jneumeth.2019.10843.

16 Optical Acoustic Monitoring of High-Voltage Transformers

D. H. Smith and J. W. Spencer

CONTENTS

16.1 INTRODUCTION

High-voltage transformers are an essential part of high-voltage electric power transmission systems. Their manifestation involves the use of wire windings immersed in electrically insulating oils, and they can be automatically switched to meet power demands using switches called *tap changers*. Continuous use over long time periods can lead to gradual deterioration, so monitoring them can assist in their maintenance and the avoidance of unnecessary power supply failures. Acoustic techniques have been used for such purposes (Cichon et al., 2011). However, monitoring such transformers is difficult because of the high voltages involved and the radio frequency active environment in which they operate making conventional electronic methods difficult to use (Warren et al., 1999; Gungor et al., 2010). The use of optical techniques based upon optical fibres provides a means for overcoming such difficulties provided suitable sensing approaches can be evolved (Jones and Spencer, 2013).

The use of optical fibre-based chromatic methods has the potential for overcoming such difficulties and for monitoring to be implemented without the need to intrude into the manufactured structure of the transformer (Deakin et al., 2014). The use of such an approach based upon optical chromaticity of acoustical signals is described, along with results obtained from in-service high-voltage transformers.

16.2 CHROMATIC MONITORING SYSTEM

Optical sensing is based upon the propagation of monochromatic light from a laser diode being propagated through a length of unjacketed multimode optical fibre attached to the outside wall of a tap changer tank of a transformer (Figure 16.1).

Acoustic signals from the transformer and tap changer mechanism produce small changes in the refractive index of the core and cladding of the fibre, which changes the propagation mode of the monochromatic light in the fibre. The output from the fibre is detected electronically and processed chromatically in the time- or frequency-transformed mode. A schematic diagram of the monitoring system is shown in Figure 16.2.

This system is an advanced form of the system mentioned in the previous chromatic monitoring book (Jones et al., 2008).

FIGURE 16.1   High-voltage transformer with optical fibre sensor attached.

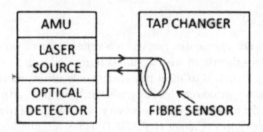

FIGURE 16.2   Schematic diagram of chromatic monitoring system.

16.3   CHROMATIC MONITORING PRINCIPLES

A typical amplitude versus time signal over a period of five minutes obtained from such an optical fibre system addressing a high-voltage transformer is shown in Figure 16.3. The acoustic signal received includes the continuous background transformer operating hum, the mechanical action of the tap changer and other acoustic events generated by the transformer.

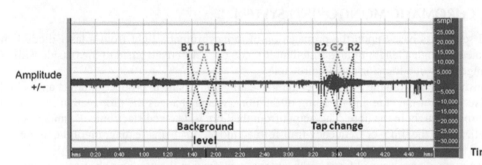

FIGURE 16.3   Typical time variation of transformer acoustic signal.

A set of chromatic filters (R, G, B) are shown being applied to the time domain signal in Figure 16.3, one set (R1, G1, B1) being applied to the background signal and a second set (R2, G2, B2) being applied to the tap changer signal. The outputs from these filters are chromatically processed (Chapter 1) to yield H, L, S values for each of the two signals. The time interval covered by each of the two sets of R, G, B filters is chosen to be 0.5 seconds (i.e., 25 cycles of the alternating current). The resulting H, L, S values may be displayed in H:L and H:S chromatic maps.

A form of secondary chromatic processing may be deployed by applying a set of R, G, B filters to parts of a longer time interval signal (e.g., 5 minutes), as shown in Figure 16.4a. The filters may then be applied to a series of time intervals during the 5-minute period, each period having values of chromatic parameters H, L, S. These H, L, S values may then be individually plotted as a function of time, as shown in Figure 16.4b. Applying a second set of chromatic filters to these primary H-S and H-L parameters yields secondary H, L, S parameters H(t), L(t), S(t) which may be used to provide additional chromatic information.

16.4 TEST RESULTS AND THEIR CHROMATIC INTERPRETATION

In situ test results have been reported (Deakin et al., 2014)) for 6 transformer units over an 11-month period during which about 13,000 tap changes occurred. These results show that the chromatic monitoring system enables two main aspects of the transformer condition to be monitored (Figure 16.3), namely the continuous operation of the transformer and the operation of the tap changers. Different forms of chromatic addressing have been described for each of these two aspects.

16.4.1 CONTINUOUS TRANSFORMER OPERATION

More details of the normal operation of a transformer (Figure 16.3) are given in Figure 16.5a and b. Figure 16.5a shows a time-varying signal obtained during a normal power-demand condition (e.g., night-time), whilst Figure 16.5b shows a signal obtained during a heavy load operation. Fourier-transformed forms of these two signals show that the heavy load condition (Figure 16.5b) is noisier than the normal load condition (Figure 16.5a). Primary chromatic processing of these two signals yields H:S and H:L chromatic maps, shown in Figure 16.6a and b. These maps show a wide distribution of points for both H:S and H:L and for the low and high power demand conditions. However, the maps suggest a more contracted distribution of both S and L for the high power load condition (Figure 16.6b).

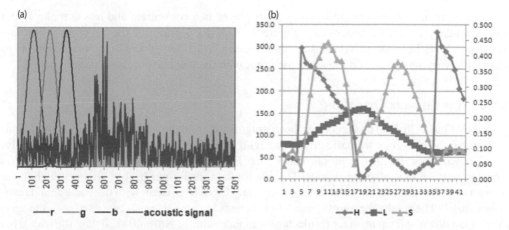

FIGURE 16.4 Secondary chromatic processing. (a) Primary chromatic processing of time varying signal. (b) Time variation of primary chromatic parameters H, L, S.

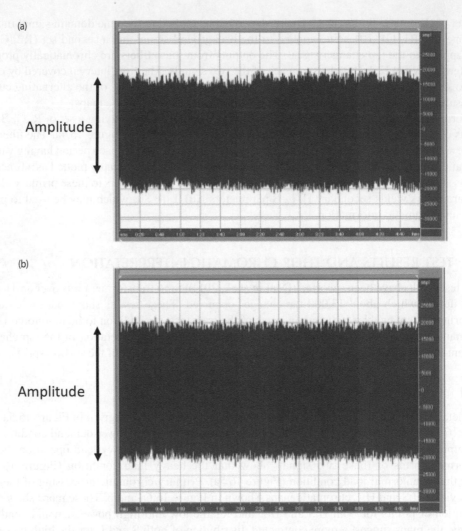

FIGURE 16.5   Time varying signals of 5-minute duration. (a) Signal with distinct harmonic peaks. (b) Signal with noisy harmonic peaks.

Trends in the transformer operation may be displayed in a convenient and less complex manner using secondary chromatic processing. Figure 16.7 shows a graph of the secondary chromatic parameters H2(H-L) versus H2(H-S). This shows results from the low and high noise levels (Figure 16.5a and b) varying linearly on the same locus but with the low noise signals having outliers.

16.4.2   TAP CHANGER OPERATION

Operation of individual tap changers may be distinguished from the background transformer signals by following the primary chromatic processing with secondary chromatic processing (Figure 16.4) to yield a secondary H(t) : L(t) polar chromatic map. Figure 16.8 presents such a H(t) : L(t) polar map for background and tap changer signals. This shows the transformer background results lying at a fixed value of H(t)=0 but with some variation in strength L(t) (consistent with the background results; Section 16.4.1) The results for four different tap changes are also shown – two each with tap up and two with tap down. All tap changer results have a higher value of H(t)=10–120 than the background results, so tap changing is chromatically distinguishable from the background. In addition, the tap down condition has substantially different values of H(t) compared with the tap up condition.

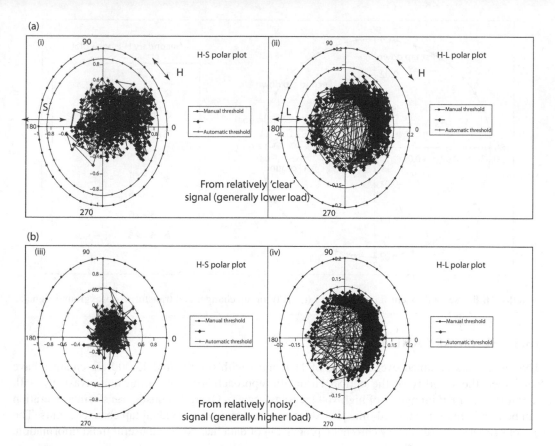

FIGURE 16.6 Primary H-S and H-L chromatic maps. (a) Distinct harmonic peaks. (b) Noisy harmonic peaks.

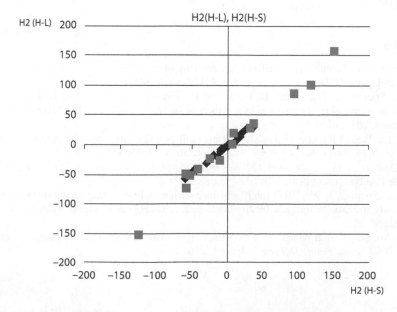

FIGURE 16.7 Secondary chromatic processed results of H2(H-L) versus H2(H-S).

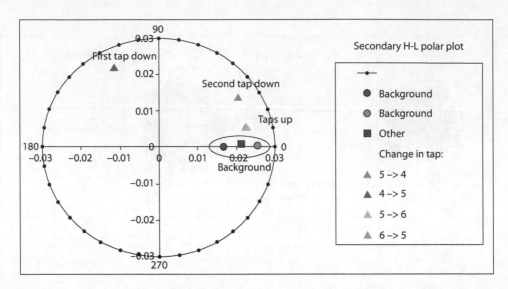

FIGURE 16.8 Secondary chromatic map of L(t) : H(t) for tap changer and transformer background signals.

16.5 SUMMARY AND OVERVIEW

The test results obtained over a period of 11 months with 6 units and 13,000 tap changes have confirmed the viability of the optical chromatic approach for monitoring non-invasively with retrofitting and RF immunity of high-voltage transformers. Overall acoustic transformer operation can be monitored for extended time periods as well as trends in individual tap changer events. The use of primary and secondary chromatic processing of data has extracted useful trend information.

ACKNOWLEDGEMENTS

The involvement and facilities provided by Electricity North West, UK and Western Power Distribution UK are acknowledged.

REFERENCES

Cichon, A., Fracz, P. and Zmarzly, D. (2011) Characteristic of acoustic signals generated by operation of on load tap changers. *ACTA phys. Polonica A*, vol. 120, no. 4, 585–588.

Deakin, A. G., Spencer, J. W., Smith, D. H., Jones, D., Johnson, N. and Jones, G. R. (2014) Chromatic optoacoustic monitoring of transformers and their onload tap changers. *IEEE Trans. On Power Delivery*, vol. 29, no. 1, 207–214.

Gungor, V. C., Lu, B. and Hancke, G. P. (2010) Opportunities and challenges of wireless sensor networks in smart grid. *IEEE Trans. Ind. Electron.*, vol. 57, no. 10, 3557–3564.

Jones, G. R., Deakin, A. G. and Spencer (2008) *Chromatic Monitoring of Complex Conditions*. CRC press (Taylor & Francis Group) ISBN 978-1-58488-988-5.

Jones, G. R. and Spencer, J. W. (2013) Intelligent monitoring of high voltage equipment with optical fiber sensors and chromatic techniques. In: *High Voltage Engineering and Testing*, Ryan, H. M. (ed.), London, U. K. Chapter 22.

Warren, C. A., Ammon, R. and Welch, G. (1999) A survey of distribution reliability measurement practices in the U S. *IEEE trans. Power Del.*, vol. 14, no. 1, 250–257.

17 Chromatic Monitoring of Arc Electrodes

Z. Wang

CONTENTS

17.1 INTRODUCTION

High-voltage circuit breakers (HVCBs) are key equipment to ensure the safety and reliable operation of the power transmission and distribution grid and therefore the reliable supply of electricity.

Switching high currents at high voltages involves mechanically separating two metallic electrodes to form an electric arc plasma which is subsequently quenched. The arcing process can cause the electrode materials to melt and evaporate, causing electrode wear which ultimately leads to the failure of the switch. There is therefore a need to monitor the extent to which the mass of the electrode is being lost during the arcing process.

The behaviour of the arc plasma and highly luminous evaporated electrode material may in principle be monitored by observing *in situ* the optical spectra of the evaporated material. However, such spectra, which evolve with time, are complex and not easily interpreted. However, chromatic analysis of such spectra can provide chromatic parameters which relate to the mass loss of an electrode during the plasma arcing process.

17.2 INSTRUMENTATION

A system for obtaining optical spectra of an arc plasma *in situ* during a current switching operation is shown in Figure 17.1. The system was mainly composed of three optical fibre sensors, one ND filter unit, one high-speed spectrometer (HSS), one photo-diode detector (PDD) unit, one digital oscilloscope, one computer and one main control unit (MCU). The material of the arcing contacts was Cu/W, and the quenching medium inside the arcing chamber was SF_6 at a pressure of 1 bar (absolute pressure) (Wang et al., 2017).

Figure 17.1 shows the electrode of an enclosed switch being addressed by the three optical fibres located at 120° to each other around the arcing contact and connected to a high-speed spectrometer. Optical signals were captured from the electrode and plasma during the current-interrupting operation.

By using this optical measuring system, the visible spectra of the arc were captured through the high-speed spectrometer. Meanwhile, the trajectory of the arc root on the plug contact surface was monitored

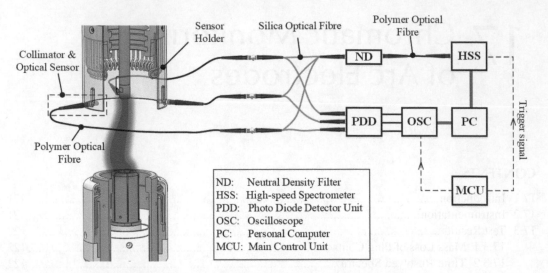

FIGURE 17.1   Schematic diagram of arc between arcing contacts plus optical measurement system (Wang, 2018).

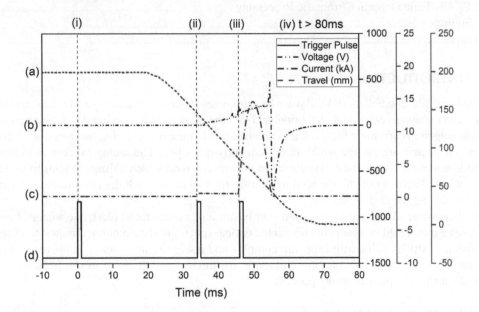

FIGURE 17.2   Time variation of trigger pulses, arc current, arc voltage and arcing contact displacement: (a) moving contact displacement, (b) arc voltage, (c) arc current, and (d) trigger pulses – (i) trigger for hydraulic mechanism, (ii) trigger for DC current, (iii) trigger for half-cycle AC current and (iv) trigger for dump ignition (Wang, 2018).

quantitatively via the photo diode detector. Figure 17.2 shows the time variation of the switch operation for a half cycle of alternating current plus the separation of the electrode to form the arc plasma.

17.3   TEST RESULTS

17.3.1   MASS LOSS OF PLUG CONTACT

Following each operation of the switch, the mass of the plug contact was mechanically measured to yield the mass lost from the electrode due to arcing. The mass loss so determined is shown in

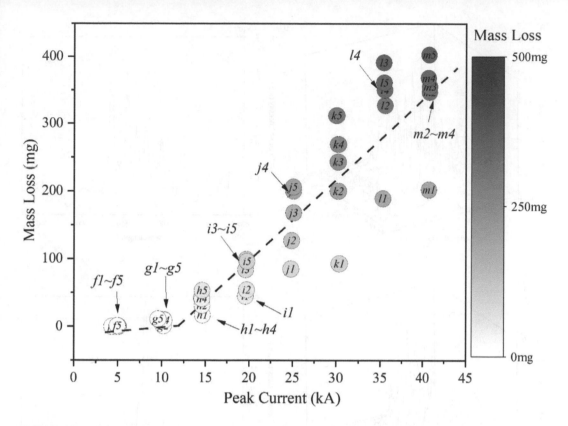

FIGURE 17.3   Directly measured mass loss from plug contact as function of peak current (Wang, 2018).

Figure 17.3 as a function of the peak alternating current (AC) and following each of five operations of the switch at eight different peak currents (5–40 kA).

17.3.2   TIME-RESOLVED SPECTRA

Typical spectra captured with the system (Figure 17.1) are shown in Figure 17.4 for different peak currents of 5 (i), 20 (ii) and 40 kA (iii). Figure 17.4a shows the time–wavelength variations of approximately ten spectra in the time window from 46 to 56 ms during the positive half cycle of the alternating current. Figure 17.4b shows the wavelength variation at the time of peak current for 5, 20 and 40 kA.

Figure 17.4a (i) shows that the arc spectra at 5 kA were mainly composed of copper line emissions from the contact surface region (i.e., 510.6, 515.3 and 521.8 nm)

Figure 17.4a (ii) and (iii) show that the profiles of the arc spectra for 20 and 40 kA (wavelength domain) were similar to each other but that their time variations were more diverse.

Figure 17.4b (i), (ii) and (iii) show more clearly the extent to which the wavelengths of the arc spectra captured at peak current time for 5, 20 and 40 kA differed.

Thus, Figure 17.4a and b illustrate the complexity of the spectra and their time variation.

17.4   CHROMATIC PROCESSING

Spectra of the form shown in Figure 17.4 were chromatically processed in the wavelength domain (primary chromatic processing) for each time interval to produce wavelength domain chromatic parameters H_w, L_w, S_w, x_w, y_w, z_w (Chapter 1; Jones et al., 2008) using the chromatic processors R_w, G_w, B_w (Figure 17.5a). Each wavelength domain chromatic parameter was then subjected to secondary chromatic processing in the time domain using time domain chromatic processors R_t, G_t, B_t

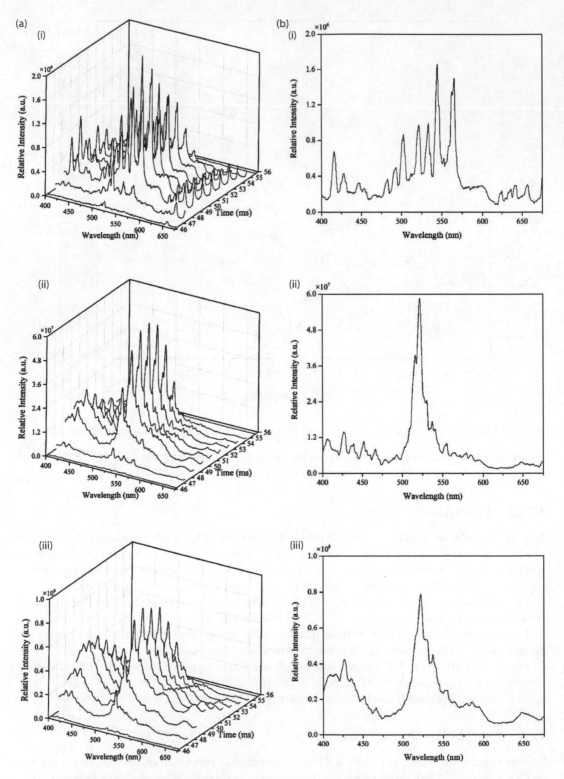

FIGURE 17.4 Optical spectra for different peak currents of 5 (i), 20 (ii) and 40 kA (iii). (a) Time–wavelength variations. (b) Wavelength variation at peak current (Wang, 2018).

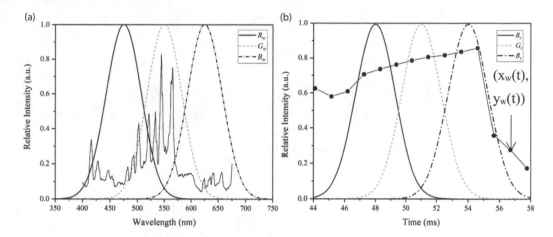

FIGURE 17.5 Chromatic processors R, G, B superimposed on wavelength and time domain signals. (a) Wavelength domain – R_w, G_w, B_w processors superimposed upon intensity versus wavelength (w) at peak current. (b)Time domain – R_t, G_t, B_t processors superimposed upon $x_w(t)$, $y_w(t)$ versus time (t).

(Figure 17.5b). This yielded a total of 36 chromatic parameters (Wang, 2018), each representing a different signal feature. Empirical test considerations were then used to select the most sensitive parameters for representing particular signal trends.

17.4.1 Wavelength Domain Chromatic Processing

A primary chromatic analysis was undertaken of the arc spectral emission at the time of peak (t_p) for each peak current investigated (e.g., Figure 17.5a). The processor outputs (Ro_w, Go_w, Bo_w) were converted into six wavelength domain chromatic parameters, $H_w(t_p)$, $L_w(t_p)$, $S_w(t_p)$, $x_w(t_p)$, $y_w(t_p)$ and $z_w(t_p)$ (Wang, 2018; Chapter 1) Two of these parameters, $y_w(t_p) = (Go_w)/3L_w(t_p)$, $x_w(t_p) = (Ro_w)/3L_w(t_p)$ [representing relative magnitudes of medium $y_w(t_p)$ and long $x_w(t_p)$ wavelength components, respectively] were used to form a chromatic map of $y_w(t_p)$ versus $x_w(t_p)$ (Figure 17.6).

The shadow level of each point in Figure 17.6 represents the mass loss of the contact (according to the shadow scale shown in Figure 17.3). This shows two different data trends which correspond to low and high currents.

At high currents (>10 kAp, tests h1–m5), the $y_w(t_p)$ parameter reduced (0.51–0.43) as the mass loss increased, with a lower trend for $x_w(t_p)$ (0.37–0.325), providing an approximate indication of mass loss. This trend corresponded to the relative optical emission in the B_w band increasing.

At low currents (5–10 kAp, tests f1–g5), the data points had a wider spread of $x_w(t_p)$ (0.7–0.26) at values of $y_w(t_p)$ between 0.475–0.5. The spectral line emission from the metallic atoms (Figure 17.4) was not predominant, implying that the vaporisation of contact material was less prominent.

17.4.2 Time Domain Chromatic Processing

The primary chromatic parameters [$x_w(t)$, $y_w(t)$] at various times (t) during a current half cycle were each addressed by three time domain chromatic parameters (R_t, G_t, B_t) as shown in Figure 17.5b. (These were empirically chosen from the possible 36 secondary chromatic [time domain] parameters; Wang, 2018.) The outputs from R_t, G_t, B_t were then used to determine the effective spectral intensity $L_{t,lw}$ and nominal time domain spread $1 - S_{t,yw}$, which were displayed on a time domain chromatic map of $L_{t,lw}$ versus $1 - S_{t,yw}$ (Figure 17.7). This map showed a linear trend locus for the two parameters which distinguished various levels of contact mass loss (Figure 17.7). Consequently, it led to two chromatic calibration graphs $1 - S_{t,yw}$ versus mass loss; $L_{t,lw}$ versus mass loss (Figure 17.8a and b) which showed that $1 - S_{t,yw}$ decreased linearly with mass loss, whilst $L_{t,lw}$ increased linearly with mass loss.

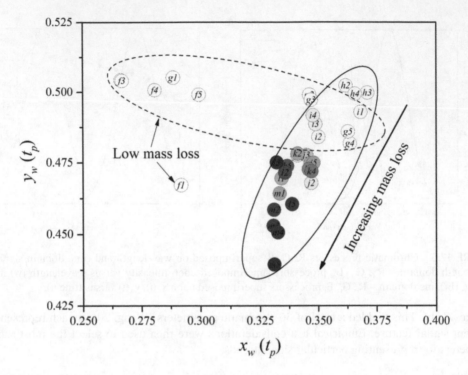

FIGURE 17.6 Primary chromatic parameters $x_w(t_p)$ versus $y_w(t_p)$ (Wang, 2018) (mass loss indicated by shadow on each point).

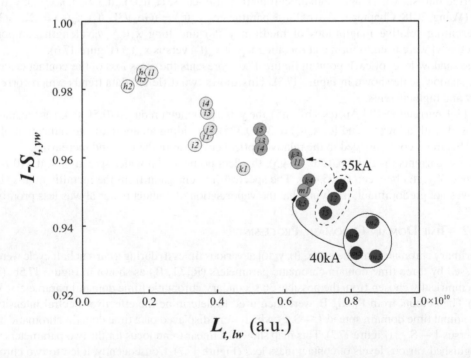

FIGURE 17.7 Secondary chromatic parameter (time domain) map of $L_{t,lw}$ versus $1 - S_{t,yw}$ (Wang, 2018).

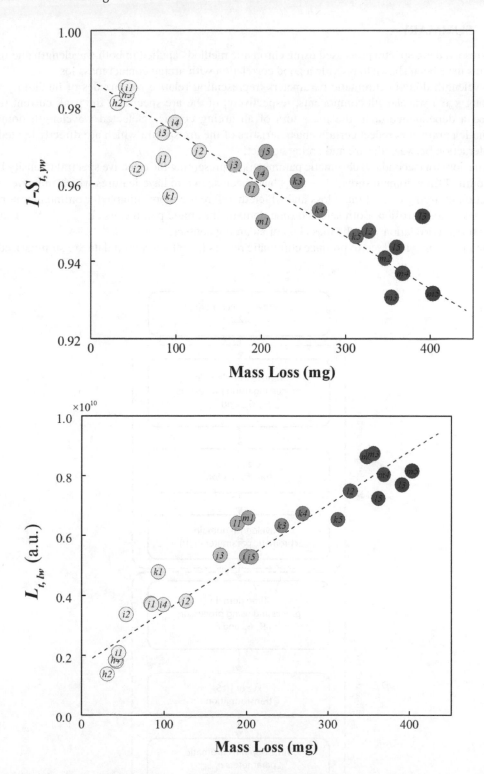

FIGURE 17.8 Secondary chromatic parameters (time domain) as a function of mass loss. (a) Effective spectral intensity $L_{t,lw}$ and (b) nominal spread $(1 - S_{t,yw})$ (Wang, 2018).

17.5   SUMMARY

Time-resolved arc spectra processed using chromatic methods applied in both wavelength and time domains have been shown to provide a good correlation with arcing contact mass loss.

Wavelength domain chromatic parameters [representing relative magnitudes of medium $y_w(t_p)$ and long $x_w(t_p)$ wavelength components, respectively] of the arc spectra at the peak current time showed a dependence upon the mass loss of an arcing contact. Selected wavelength domain chromatic parameters reflect certain characteristics of the arc spectra which are directly related to the interaction between the arc and arcing contact.

Time domain (secondary) chromatic parameters (representing the effective spectral intensity $L_{t,lw}$ and nominal time domain spread $1 - S_{t,yw}$) have been shown to have features which improved the prediction of arcing contact mass loss and which therefore are more suitable for online monitoring of arcing contact erosion. Both selected time domain chromatic parameters ($L_{t,lw}$, $1 - S_{t,yw}$) show strong linear correlation with the mass loss of an arcing contact.

The procedures followed to produce chromatic results from the spectral data are summarised in Figure 17.9.

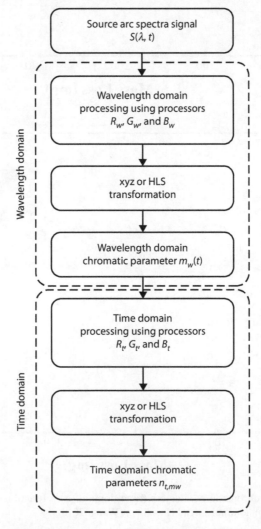

FIGURE 17.9   Flow chart for processing spectral data using primary and secondary chromatic methods (Wang, 2018).

REFERENCES

Jones, G. R., Deakin A. G. and Spencer, J. W. (2008) *Chromatic Monitoring of Complex Conditions.* CRC Press, ISBN 978-1-58488-988-5.

Wang, Z. (2018). Study on the Mechanisms and Prediction Methods of Arcing Contact Erosion of High-Voltage SF6 Circuit Breaker, Ph.D. Thesis, University of Liverpool.

Wang, Z., Jones, G. R., Spencer, J. W., Wang, X. and Rong, M. (2017) Spectroscopic on Line Monitoring of Cu/W Contacts Erosion in HVCBs Using Optical Fibre Based Sensors and Chromatic Methodology. *Sensors*, 17 (3), 519.

18 Chromatic Monitoring of Alternating Current Partial Discharges

M. Ragaa and G. R. Jones

CONTENTS

18.1 INTRODUCTION

Electrical insulation is required for preventing short-circuit faults occurring in electrical equipment and systems (e.g., Ragaa, 2012). In addition to normal electrical stress, electrical insulation is often exposed to abnormal electrical stress such as higher-frequency harmonics and impulse voltages produced from a variety of causes, for example, due to switching, lightning voltage impulses and so on. (Ariastina and Blackburn, 2000). Such effects can transiently produce short-duration, unsustainable electrical discharges which are known as partial discharges (PDs). Repetitive exposure to such discharges can eventually lead to undesirable insulation failure. Consequently, monitoring such partial discharges unobtrusively provides a means for tracking any impending failure of the electrical insulation which can have an undesirable impact upon equipment and system operation. Partial discharges can be of different forms. Figure 18.1a shows a PD known as a "tree" formed along the surface of an electrical insulator (Cavallini et al., 2004). Partial discharges may also be formed within solid electrical insulators (Figure 18.1b) or within liquids such as electrically insulating oils (Figure 18.1c) (Ariastina and Blackburn, 2000).

Electromagnetic emissions which accompany partial discharges can contain useful information about the condition of the electrical insulation and the likelihood of possible insulation failure. Chromatic analysis of such emissions may provide an insight into the partial discharging, such as its severity, frequency and possible consequences. This chapter describes the application of chromatic analysis of partial discharge signals for interpreting implications of the partial discharging.

18.2 DATA ANALYSIS

The PD signals are in the form of two high-frequency pulses which occur every half cycle of an alternating current (AC) cycle. The amplitudes of each of the two pulses are positive and negative, respectively, as shown in Figure 18.2.

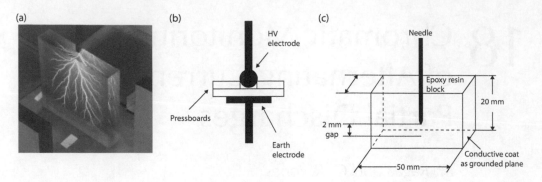

FIGURE 18.1 Examples of partial discharges (PD). (a) Surface discharge (tree). (b) Point–plane gap (Lai et al., 2008). (c) Sphere in liquid (Ariastina and Blackburn, 2000; Ragaa, 2012).

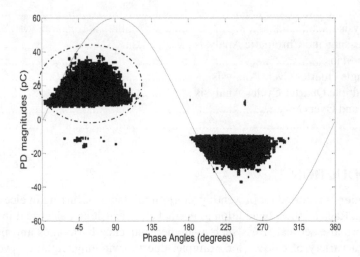

FIGURE 18.2 Example of the two components of a partial discharge signal (Lai et al., 2008).

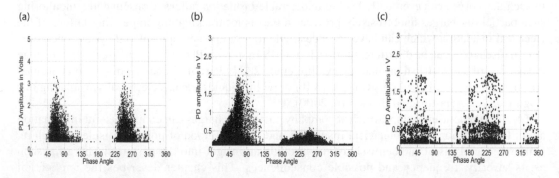

FIGURE 18.3 Examples of charge amplitude: phase angle signals produced by different PDs for 50 Hz voltage waves. (a) Internal (21 kV). (b) Surface (6 kV). (c) Floating in oil (23 kV) (Ragaa, 2012).

Examples of signals produced in an AC system by different types of partial discharges (with the second half cycle rectified) are shown in Figure 18.3. Figure 18.3a–c are, respectively, for internal, surface and floating oil discharges (Ariastina and Blackburn, 2000; Cavallini et al., 2004; Lai et al., 2008; Ragaa et al., 2012).

Details about the various partial discharges are given in Table 18.1.

TABLE 18.1

Sources of PD Data

Source	PD Type	PD Trend
Lai et al. (2008)	Point–plane	Time to BD
Ariastina and Blackburn (2000)	Void in liquid	Test voltage
Cavallini et al. (2004)	Treeing in cable	Tree time

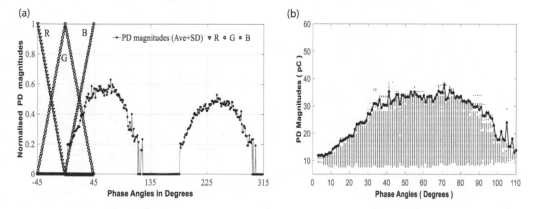

FIGURE 18.4 PD data conditioning for chromatic analysis. (a) Normalised amplitude per phase degree and chromatic processors (R, G, B) addressing first quarter cycle (Ragaa et al., 2012). (b) Digitised PD magnitude, first half cycle (signal; Lai et al., 2008) (boundary → average level + 1.7 standard deviations).

18.3 DATA PROCESSING AND CHROMATIC ANALYSIS

A positive half cycle PD signal in the form of pC amplitude as a function of phase angle accumulated over 30 cycles of AC voltage at 50 Hz is digitised (Figure 18.4).

An envelope of the digitised signal is determined as the mean plus 1.7 standard deviations. The resulting envelope is addressed by three non-orthogonal processors R, G, B, as shown in Figure 18.4a. The phase extent of the three processor responses may be chosen to cover various quarter cycles of the positive and negative cycles of the waveform. Figure 18.4a shows the first quarter cycle being addressed by R_1, G_1, B_1, whose outputs are transformed into signal-quantifying chromatic parameters (x_1, y_1, z_1, L_1, S_1, H_1) which are used to form appropriate chromatic maps. Two such maps are utilised: (a) $x_1 : y_1$ map to represent the effective spread of PD activity for the first quarter cycle and (b) $L/(1 - S) : z_1$ to represent an effective amplitude versus the relative B component. A summary of the chromatic processing procedures is given in Figure 18.5.

A similar analysis may be used to provide additional parameters for the second, third and fourth quarter cycles (e.g., x_2, y_2; x_3, y_3; x_4, y_4 etc).

18.4 CHROMATIC MAPS

18.4.1 Single Quarter Cycle Analysis

An $x_1 : y_1 : z_1$ chromatic map compares the relative magnitudes of the PD activity within the three sectors of the first quarter cycle of the PD-inducing AC wave. As a guide to interpreting results in such a map, the chromatic response to some simple pulse leading-edge geometries may be

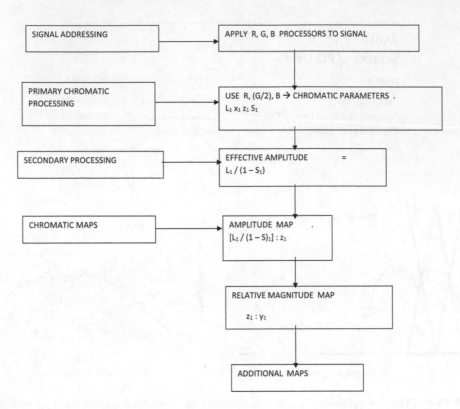

FIGURE 18.5 Summary of chromatic processing procedures.

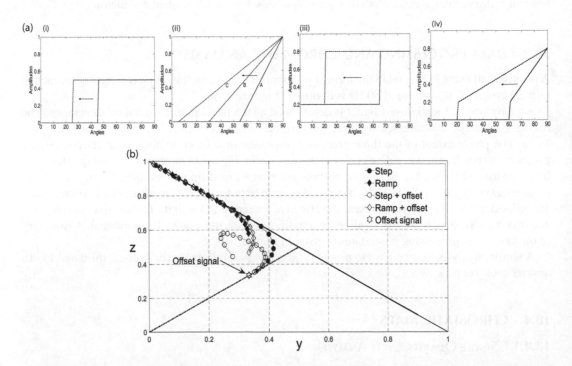

FIGURE 18.6 Chromatic transformation of some simple signal envelopes: (a) Amplitude versus phase angle for the first 90° quarter cycle of a power frequency wave. (i) Step function, (ii) ramp, (iii) step + offset, (iv) ramp + offset (Ragaa, 2012). (b) z versus y chromatic map of each signal spreading across the first quarter cycle (Ragaa, 2012).

followed on such a $z_1 : y_1$ map. Examples of such leading-edge geometries include a step change [Figure 18.6a (i)], a ramp change [Figure 18.6a (ii)], a step change plus offset [Figure 18.6a (iii)] and a ramp change plus offset [Figure 18.6a (iv)]. Figure 18.6b shows a $z_1 : y_1$ chromatic map with the chromatic signatures of these four examples superimposed. These show that a step change is initially identified as being located on the locus $z_1 + y_1 = 1$ and shifted along this locus towards $z_1 = y_1$ before departing off the locus towards the equal value point $z_1 = y_1 = x_1 = 0.33$. The ramp change initially follows the same locus ($z_1 + y_1 = 1$) but deviates from this locus sooner than the step change. The step plus offset variation is displaced from the $z_1 + y_1 = 1$ locus before tending towards the equal value point of 0.33. The ramp and offset results follow a similar path to the ramp results. Thus, these results indicate that a preliminary indication of the form of leading-edge geometry may be obtained from a $z_1 : y_1$ chromatic map along with the extent to which the signal extends along the AC quarter cycle.

Results for the various types of partial discharges shown in Figure 18.3a are given in the $z_1 : y_1$ chromatic maps of Figure 18.7a (i), (ii), (iii). The results of Ariastina for different PD-producing AC voltage shift along the locus $z_1 + y_1 = 1$ towards higher y_1 values before tending to the equal value coordinate (0.33, 0.33). This indicates the PD envelope expands within the quarter cycle until ultimately, full electrical breakdown occurs. For a fixed AC voltage and point plane geometry (Lai et al., 2008), the PD signal shifts along the locus $z_1 + y_1 = 1$ before moving off the locus sooner, just before full electrical breakdown.

Although the $x_1 : y_1 : z_1$ chromatic maps provide valuable indications of PD activity, they do not directly give an indication of the strength of such activity. To obtain such information, recourse may be made to the $L/(1 - S) : z_1$ map, with which $L/(1 - S)$ provides an indication of the effective amplitude of the PD. Such a map is shown in Figure 18.7b for the test results shown in Figure 18.7a. This map shows that for the increasing AC voltage producing the PD (Ariastina and Blackburn, 2000), the effective amplitude of the PD increases as the discharge tends towards breakdown, providing additional confirmation of the trend suggested by the $x_1 : y_1 : z_1$ chromatic map. The data of Cavallini et al. (2004) also shows an increase in the effective PD amplitude as electrical breakdown is approached, albeit at a lower level of the z1 relative spread. In the case of the Lai et al. (2008) data, the pre-breakdown effective amplitude is considerably higher than for the Ariastina and Blackburn (2000) and Cavallini et al. (2004) data.

18.4.2 Multiple Quarter Cycles Analysis

Additional information about the form of the partial discharge pulse may be obtained by comparing other quarter cycle chromatic parameters (e.g., x_2 from the second half cycle, z_3 from the third half cycle etc.) for the rectified positive and negative AC half cycles (Figures 18.2 and 18.4a). For example, a chromatic map of z_1 versus z_3 (Figure 18.8a; Ragaa, 2012) indicates $z_1 \sim z_3$, implying no substantial polarity effects on the distribution. A chromatic map of z_1 versus x_2 (Figure 18.8b; Ragaa, 2012) provides an indication of the degree of symmetry in the PD distribution about the peak AC voltage of the first half cycle. Test results lying on either side of the locus $z_1 = x_2$ indicate an asymmetric distribution, the degree of asymmetry being indicated by the deviation from the $z_1 = x_2$ locus.

The various chromatic maps described enable information about a partial discharge activity to be derived as indicated on the flow chart in Figure 18.9. The $L_1/(1 - S_1) : z_1$ map indicates the amplitude of the PD and its extent z_1. As such, it can indicate an approach to electrical breakdown plus the kind of partial discharge occurring. Confirmation is provided by details of the PD during the first quarter cycle indicated by the $x_1 - y_1$ chromatic map plus the $z_1 - z_3$ and $z_1 - x_2$ maps.

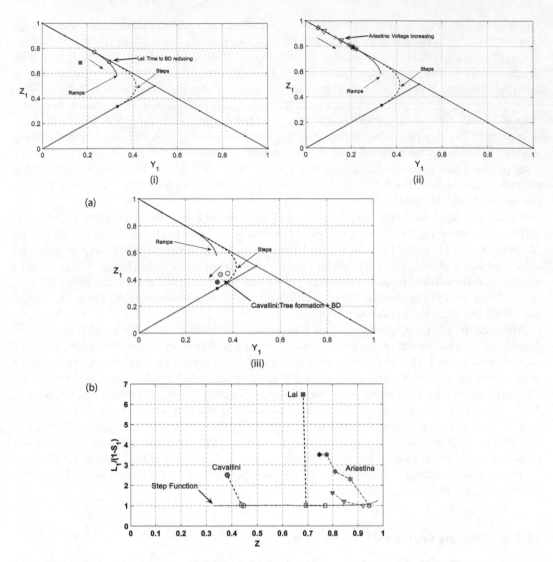

FIGURE 18.7 Chromatic maps of different types of partial discharges (first quarter cycle ($-45°$ to $+45°$)) (Ragaa, 2012) (i) Point–plane gap (Lai et al., 2008). (ii) Sphere in liquid. (iii) Treeing in cable. (a) z_1 versus y_1, (b) effective amplitude $L_1/(1 - S_1)$ versus relative spread, z_1, a voltage wave for each data set of Table 18.1.

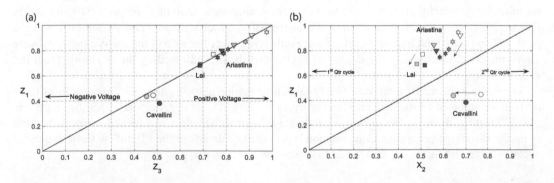

FIGURE 18.8 Multiple quarter cycle chromatic maps (Ragaa, 2012). (a) $z_1 = B_1/3L_1$ versus $z_3 = B_3/3L_3$, voltage polarity effect. (b) $z_1 = B_1/3L_1$ versus $x_2 = R_2/3L_2$, pulse symmetry effect.

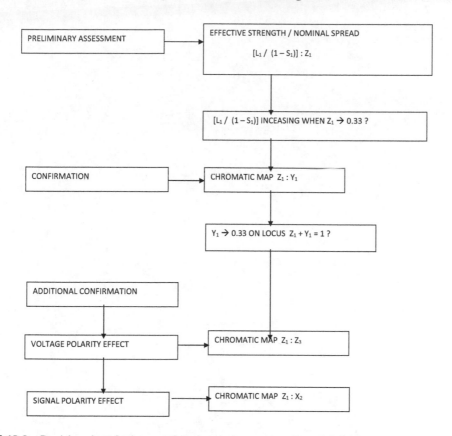

FIGURE 18.9 Decision chart for interpreting chromatic mapping of partial discharges.

18.5 SUMMARY AND OVERVIEW

A chromatic analysis of partial discharge signals produced during the leading quarter cycle of an alternating current has been processed chromatically. It has enabled the progression of the PD towards full electrical breakdown to be indicated. The type of PD producing a particular signal may be identified in various chromatic maps. This can assist in tracing the source of the PD before the occurrence of full electrical insulation failure. Examination of the chromatic maps for other quarter cycle PDs provides additional confirmation of the PD activity.

REFERENCES

Ariastina, W. G., and Blackburn, T. R. (2000) Characteristics of partial discharge in oil impregnated insulation using different test voltage frequencies. *IEEE International Conference on Properties and Applications of Dielectric Materials*, pp. 487–492.

Cavallini, A., Conti, M., Montanari, G., Arlotti, C., and Contin, A. (2004) PD inference for the early detection of electrical treeing in insulation systems. *IEEE Transactions on Dielectrics and Electrical Insulation*, 11(4), 724–735.

Lai, K. X., Lohrasby, A., Phung, B. T., and Blackburn, T. R. (2008) Partial discharge of electrical trees prior to breakdown. *Int. Symposium on Electrical Insulating Materials*, pp. 649–652.

Ragaa, M. (2012) Chromatic monitoring of partial discharge signals. Ph.D. Thesis, University of Liverpool, UK.

Ragaa, M., Jones, G. R., Deakin, A. G., and Spencer, J. W. (2012) Partial discharge monitoring using chromatic maps. *Proc. 19th Gas Discharge Conference*, pp. 742–745.

19 Transmission Line Fault Diagnosis Using Chromatic Monitoring

Ziyad S. D. Almajali

CONTENTS

19.1 INTRODUCTION

In electrical power systems, the overhead transmission line (Figure 19.1a) is utilised for facilitating power delivery of electrical energy to consumers despite the long distance separating it from its resources. Standard three phase parallel transmission lines increase the power transmission capability. A schematic diagram of parallel three phase power lines is shown in Figure 19.1b. Compared to the generation and distribution parts of the power system, high failure probability is recorded in the transmission part despite the high standard measures in the design and material selection.

In order to ensure a continuous and reliable supply of electric power, any faults occurring in such systems need to be detected urgently and the location and type of fault identified. Monitoring can only be done conveniently at both ends of the transmission line and the fault location and type determined from the source and receiving end signals.

Different faults may occur with large possibilities of different locations throughout the line. Monitoring is made more difficult because the voltage and current waveforms of each of the three phases may show complex variations upon fault occurrence. The variation depends on the severity of the fault and how close the fault is to the measurement points. Such complexity makes it difficult to directly interpret the waveform variation into useful information about the type of the fault which is causing it and the location.

The unknown and unpredictable fault parameters promote the use of artificial intelligence–based methods such as fuzzy logic (Ferrero et al., 1994; Kumar et al., 1999) and neural networks (NNs) (Mohamed and Rao, 1995; Aggarwal et al., 2012) in addition to knowledge-based methods of wavelet transforms (Shaik and Pulipaka, 2015) and a combination of various methods (Reddy and Mohanta, 2007; Pothisarn and Ngaopitakkul, 2010; He et al., 2014).

Chromatic monitoring techniques have also been considered using computer-based simulation packages (Almajali et al., 2013, 2014). This has involved applying three chromatic processors to 60-Hz alternating current waveforms and extracting values of the resulting chromatic parameters to yield the required information.

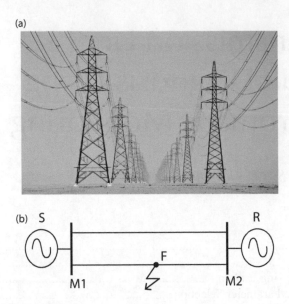

FIGURE 19.1   Three phase electric power transmission lines. (a) Power system transmission lines. (b) Schematic diagram of parallel three phase power lines and a fault (F) [M1 = Source (S), M2 = Receiver (R)].

19.2   METHODOLOGY

19.2.1   PRE-PROCESSING

The available data for fault detection are in the form of two groups of sinusoidal waveforms, distinguished by the collecting ends of the transmission lines (sender or receiver) from which they have been obtained. Each group of waveforms from each end contain another two groups, three currents and three voltages from the three current phases (I_a, I_b, I_c; Figure 19.1b). Chromatic processing involves applying three chromatic processors (R, G, B) to these alternating current waveforms and extracting values of the resulting chromatic parameters (H, L, S) (Chapter 1).

In principle, all the available signals should be monitored, which would require a comprehensive and complex diagnostic system. However, a variation in the waveform characteristics due to a fault occurs not only in the faulty phase but also in the remaining phases, albeit reduced (Anderson, 1995). Therefore, the symmetrical components method is utilised as a pre-processing step to transform the three phase waveforms of both current and voltage into new different sets of waveforms. Thus, monitoring any one of the three symmetrical components avoids the need to monitor all three current components and reduces the calculation burden.

Figure 19.2a shows three typical alternating current (AC) waveforms corresponding to each of the three phase currents (I_a, I_b, I_c) at the receiving end of a transmission line. Phase I_b suffered a fault at a time of 33.5 ms.

The positive (I_{a1}), negative (I_{a2}) and zero (I_{a0}) sequence components are calculated as follows (Fortescue, 1918).

$$I_{a1} = [1/3][I_a + \alpha \cdot I_b + \alpha^2 \cdot I_c] \tag{19.1}$$

$$I_{a2} = [1/3][I_a + \alpha^2 \cdot I_b + \alpha \cdot I_c] \tag{19.2}$$

$$I_{a0} = [1/3][I_a + I_b + I_c] \tag{19.3}$$

where α is a unit vector at an angle of 120°. $\alpha = 1\angle 120°$ and $\alpha^2 = 1\angle 240°$.

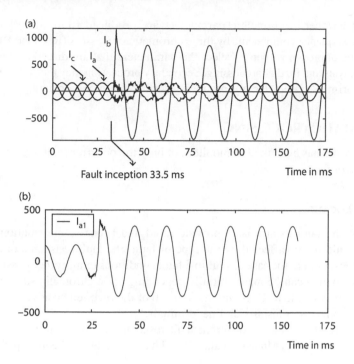

FIGURE 19.2 Three phase current waveforms with fault. (a) I_a, I_b, I_c. (b) Positive symmetrical component (I_{a1}).

An example of the positive sequence component (I_{a1}) for I_a, I_b, I_c (Figure 19.2a) is shown in Figure 19.2b.

19.2.2 CHROMATIC PARAMETER SELECTION

The chromatic monitoring method is based on the continuous monitoring of vital signals from the power system (current and voltage waveforms) in the time domain. The process is applied to a series of time windows, each of a single cycle time duration. HLS chromatic parameters are calculated for each waveform from the outputs of the processors R, G, B (Chapter 1). An example illustrating the RGB filtering implementations on a fully rectified post-fault cycle of a positive sequence waveform is shown in Figure 19.3.

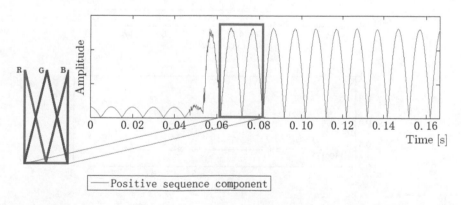

FIGURE 19.3 An example of RGB processor application to a post-fault cycle of a rectified positive sequence component waveform.

A substantial number of chromatic parameters are produced from addressing the available waveforms of the sequence components by three chromatic processors (R, G and B) (Equations 19.1 through 19.3). The approach is then to select the parameters that are directly affected by the fault conditions for analysis and further processing. Secondary processing may then be used for connecting different sets of primary parameters for producing further useful parameters.

19.3 CHROMATIC PROCESSING

Chromatic processing has been used to monitor not only the occurrence of a fault but also the fault location and type of fault.

19.3.1 FAULT LOCATION

The objective of the fault locator is to provide rapid and accurate information about the fault location. This should not be affected by the type of fault, the fault resistance or any surrounding condition variations, even in the case of multiple fault conditions or any other possible disturbances. Fault location estimation can be made initially depending upon chromatic strength parameter (L) evaluation from the available R, G, B outputs for each of the rectified positive current symmetrical components and from both transmission line terminals.

To overcome any effects due to different fault resistances and so on, a secondary chromatic parameter (L_{RS}) is introduced (Almajali et al., 2013). This parameter has been empirically determined

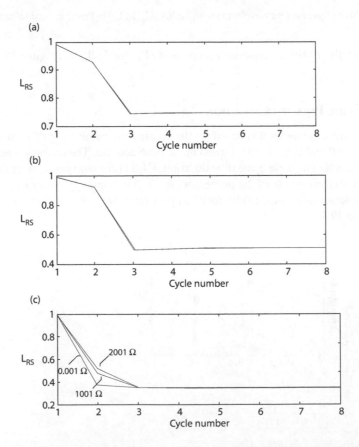

FIGURE 19.4 Variation of L_{RS} over several cycles close to a fault. (a) At 75% of the line length, (b) at 50% of the line length, (c) at 35% of the line length with different fault resistances (Almajali, 2015).

as the relative strength difference between the source and receiver strengths $(L_r - L_s)$ with respect to the sum of the source and receiver strengths $(L_r + L_s)$. The parameter is given by Equation 19.4.

$$L_{RS} = (L_r - L_s)/(L_r + L_s) \qquad (19.4)$$

An example of the variation of LRS with time over several cycles close to a fault inception is given in Figure 19.4 for faults at different locations of the line.

The LRS value in Figure 19.4a reduces from unity at fault inception (cycle 1) to 0.75 at cycle 3, indicative of the fractional location of the fault halfway along the line length. LRS for faults at other line locations (0.5 of line length) is shown in Figure 19.4b, while Figure 19.4c illustrates how LRS transition is also independent of the fault electrical resistance by an example of a fault at 0.35 of line length but with different fault resistances (0.001, 100 and 200 Ω). The method indicates a fault location to an accuracy of 1.9%, and this value is not affected by the fault electrical resistance or the position of the fault (Almajali et al., 2013).

Figure 19.5 shows a flowchart for the complete process of the locator procedure, which involves required waveforms, in addition to the stages of data pre-processing, primary processing and secondary chromatic processing.

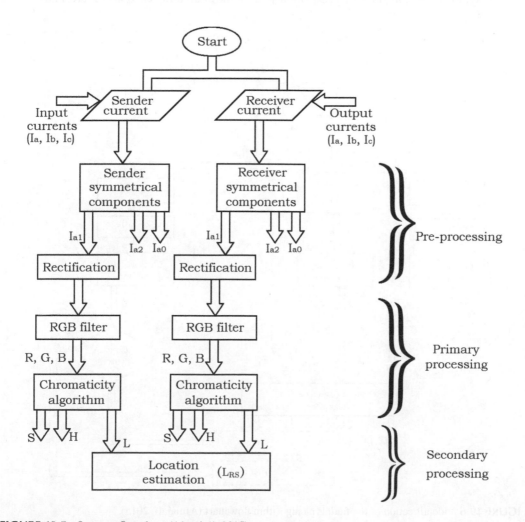

FIGURE 19.5 Locator flowchart (Almajali, 2015).

19.3.2 IDENTIFICATION OF FAULT TYPE

An indication of the fault type can be made chromatically to various levels of detail depending upon the chromatic parameters used. R,G,B outputs are available for each of the three current phases (I_a, I_b, I_c). The R, G, B outputs are processed to provide several H, L, S chromatic parameters. Figure 19.6 shows a flowchart which indicates the various combination of chromatic outputs and parameters which have been shown to provide various levels of fault types (Almajali et al., 2014). Physically relevant fault types may be classified as belonging to one of three groups:

a. Faults between two of the three phases (AB, AC, BC)
b. Faults between each phase and ground (AG, BG, CG)
c. Faults between pairs of phases and ground (ABG, ACG, BCG)

where A, B, G refer to current phases I_a, I_b, I_c, respectively and G refers to the ground involvement. The flow diagram indicates which chromatic parameters may be used to distinguish between the different faults.

An example of the fault type distinction which can be achieved is the use of the H value of the negative symmetrical component I_{a2} (Equation 19.2) of the alternating current (Figure 19.6).

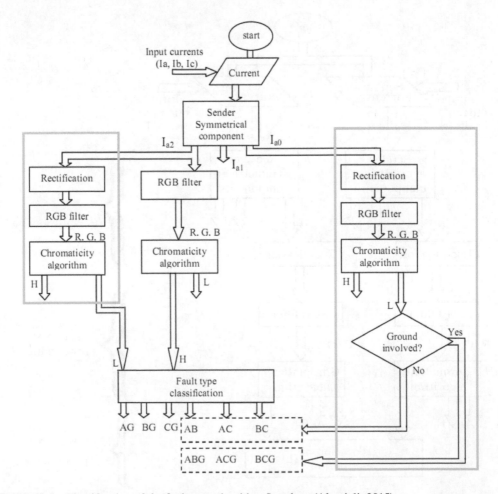

FIGURE 19.6 Identification of the fault type algorithm flowchart (Almajali, 2015).

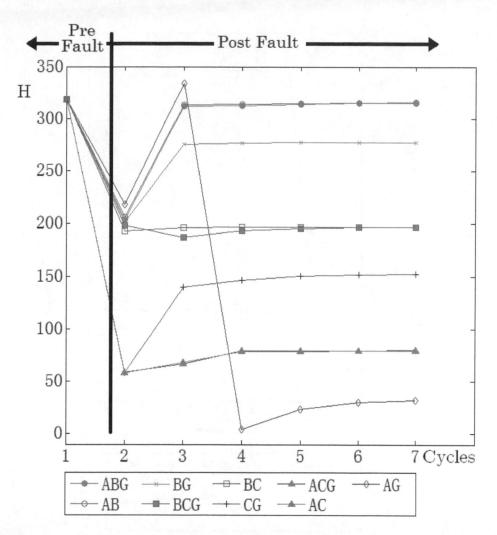

FIGURE 19.7 Chromatic analysis to indicate fault type; H from negative symmetrical component (I_{a2}) (Almajali, 2015).

Figure 19.7 shows the H parameter for I_{a2} as a function of number of cycles for different types of faults at 50% of the transmission line length. The fault types cover all possible range of faults. The results show that various faults may be grouped according to different ranges of H but with some ambiguities; that is, the line-ground faults (ABG, ACG, BCG) are not distinguishable from the line-only faults (AB, AC, BC), and ambiguities can occur at the boundaries between each of the six ranges.

Limitations with the H parameter-based fault type identifier may be addressed through the use of the L parameter for the rectified zero symmetrical component I_{a0} (Figure 19.6).

Figure 19.8 shows the L parameter for the rectified I_{a0} as a function of number of cycles close to a fault of different types. This shows that the healthy condition together with the nine different faults form three clusters of L parameter levels. However, the non-ground and lines to ground faults are distinguishable by their different values of the chromatic L parameter.

If the H parameter information is combined with the L parameter information, ambiguities can be removed to provide a higher level of discrimination, as shown in the L of I_{a0} versus H of I_{a2} graph of Figure 19.9.

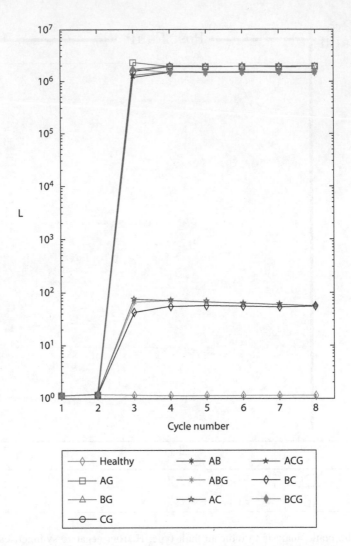

FIGURE 19.8 Chromatic analysis to indicate ground presence; L from rectified zero sequence component (I_{a0}) (fault resistance = 0.001 Ω, fault location = 50%) (Almajali, 2015).

19.4 SUMMARY AND OVERVIEW

Chromatic techniques can be deployed for overhead transmission line fault diagnosis by identifying the fault type and location estimation. A combination of the signal strengths (Equation 19.4) of the rectified positive sequence components (I_{a1}; Equation 19.1) for the line at the source (L_s) and receiver (L_r) ends of a three phase power transmission line can identify the location of a fault along the line regardless of the electrical resistance of the fault (Figure 19.4).

The use of the chromatic parameter H of the negative sequence component (I_{a2}; Equation 19.2), along with the strength L for the rectified zero sequence component (I_{a0}) of an AC waveform enables various types of faults (single line to ground, line to line, line to ground) to be distinguished (Figure 19.9).

The chromatic approach provides a high level of transparency and traceability in monitoring electrical transmission line faults. Further development of the chromatic parameters available can lead to further fault discrimination capabilities.

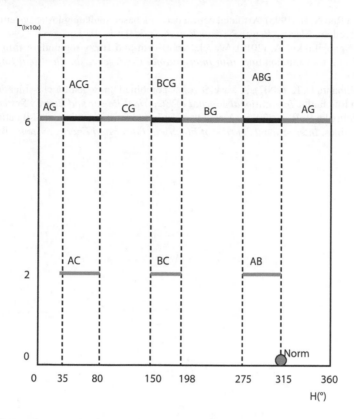

FIGURE 19.9 Chromatic map of $L(I_{a0})$ versus $H(I_{a1})$ with clusters of different faults (AG, BG, CG and AC/ACG, BC/BCG, AB/ABG) plus normal condition (Norm) (Almajali, 2015).

REFERENCES

Aggarwal R. K., Blond S. L., Beaumont P., Baber G., Kawano F., and Miura S. (2012). High frequency fault location method for transmission lines based on artificial neural network and genetic algorithm using current signals only. *IET Conference Publications*, volume 2012, University of Bath.

Almajali Z. S. (2015). Fault Diagnosis for Transmission Lines Using Chromatic Processing, Ph.D. thesis University of Liverpool.

Almajali Z. S., Spencer J.W., and Jones G.R. (2013). Fault Locator for a Parallel Transmission Line Using Chromatic Processing. *The 8th Jordanian International Electrical and Electronics Engineering Conference*, (JIEEEC 2013). Amman, Jordan.

Almajali Z. S., Spencer J.W., and Jones G.R. (2014). Asymmetrical Fault Classifier for a Parallel Transmission Line Using Chromatic Processing. *The 7th IET International Conference on Power Electronics, Machines and Drives (PEMD 2014)*. Manchester, UK.

Anderson P. M. (1995). *Analysis of Faulted Power Systems [Electronic Book]*. IEEE Press Series on Power Engineering, Piscataway, NJ.

Ferrero A., Sangiovanni S., and Zappitelli E. (1994). A fuzzy-set approach to fault-type identification in digital relaying. Transmission and Distribution Conference, *Proceedings of the 1994 IEEE Power Engineering Society*, pp. 269–275.

Fortescue C. L. (1918). Method of symmetrical co-ordinates applied to the solution of polyphase networks. *Proceedings of the American Institute of Electrical Engineers*, 37(6): 629–716.

He Z., Lin S., Deng Y. J., Li X. P., and Qian Q. Q. (2014). A rough membership neural network approach for fault classification in transmission lines. *International Journal of Electrical Power and Energy Systems*, 61: 429–439.

Kumar P., Jamil M., Thomas M. S., and Moinuddin (1999). Fuzzy approach to fault classification for transmission line protection. TENCON 99. *Proceedings of the IEEE Region 10 Conference*, volume 2, pp. 1046–1050.

Mohamed E. A. and Rao N. D. (1995). Artificial neural network based fault diagnostic system for electric power distribution feeders. *Electric Power Systems Research*, 35(1): 1–10.

Pothisarn C. and Ngaopitakkul A. (2010). Wavelet transform and fuzzy logic algorithm for fault location on double circuit transmission line. *16th International Conference on Electrical Engineering (ICEE)*, Busan, Korea.

Reddy M. J. and Mohanta D. K. (2007). A wavelet-fuzzy combined approach for classification and location of transmission line faults. *International Journal of Electrical Power and Energy Systems*, 29: 669–678.

Shaik A. G. and Pulipaka R. R. V. (2015). A new wavelet based fault detection, classification and location in transmission lines. *International Journal of Electrical Power and Energy Systems*, 64: 35–40.

20 Chromatic Analysis of High-Voltage Transformer Oils Data

E. Elzagzoug and G. R. Jones

CONTENTS

20.1 INTRODUCTION

To ensure the provision of reliable and uninterrupted electric power supplies, it is essential to routinely monitor the condition of installed high-voltage power system components such as transformers (Chiesa, 2010). A number of methods are conventionally used by the industry for checking the condition of such high-voltage transformers. These include the analysis of gases dissolved in the electrically insulating oils referred to as Dissolved Gas Analysis (DGA). The relative amounts of different gases present give an indication of possible developing faults such as overheating, electrical breakdown and so on (Duval diagram) (Wang, 2000). In addition, laboratory testing of the oil for the amount of water and acid present along with the dielectric strength of the oil can be used as indicators of incipient faults. The colour of the oil via colour index (CI) is used as a further indication of abnormalities (Golden et al., 1995).

Interpretation of the overall implications of this complex array of data is difficult and can be contradictory. However, processing this complex array of data can be achieved chromatically to quantify the implication of combining the data array and to provide a traceable route to indicate the relative magnitudes of the various effects (Jones et al., 2008; Looe and Zulfadhly, 2015).

A chromatic procedure for combining the various test results obtained with conventional techniques used in the industry may be based upon the use of the following chromatic methods:

1. Primary chromatic analysis of the level of each gas dissolved in the transformer oil using the discrete chromatic approach (Chapter 1).
2. Primary chromatic analysis of the water content, acidity and dielectric strength of an oil sample using the discrete chromatic approach (Chapter 1)
3. Chromatic interpretation of colour index data for linking into the broader chromatic approach.
4. Secondary chromatic analysis applied to dominant chromatic parameters from each of the previous primary chromatic analyses.

The whole procedure may be summarised in a chromatic flowchart.

20.2 PRIMARY CHROMATIC ANALYSIS OF DISSOLVED GASES

Seven gases which are commonly monitored in dissolved gases tests are CH4, H2, C2H6, C2H2, CO2, CO and C2H4 (Elzagzoug et al., 2014). These gases are normally tested and the results normalised in accordance with international regulatory bodies (International Electrotechnical Commision [IEC], International Council on Large Electrical Systems [CIGRE]) guidelines (Leibfried et al., 2002). The gases may be divided into three groups according to the physical conditions which produce them (Aragon-Patil et al., 2007) (Figure 20.1), that is, electrical discharging within the transformer, excessive temperature rises (which could cause for example cellulose decomposition >700C) and lower temperature heating (<700C). The dominant gases in each of the three groups are CH4, C2H2 and C2H4 (Figure 20.1). Each gas forms a discrete signal element (Chapter 1), and the three groups of gases are addressed by three non-orthogonal processors (R1, G1, B1) with triangular responses (Figure 20.2). The outputs from the processors are chromatically processed (Chapter 1) to produce an effective magnitude parameter L(DGA) and chromatic map parameters X(DGA), Y(DGA), Z(DGA) defined by the equations (Chapter 1), that is

$$X(DGA) = R(DGA)/3L(DGA)$$

$$Y(DGA) = G(DGA)/3L(DGA)$$

$$Z(DGA) = B(DGA)/3L(DGA)$$

$$L(DGA) = [(R(DGA) + G(DGA) + B(DGA)]/3$$

X(DGA) represents the C2H4 dominant gas group, Y(DGA) the C2H2 group and Z(DGA) the CH4 group (Chapter 2). The X(DGA), Y(DGA), Z(DGA) chromatic map is shown in Figure 20.3a to indicate which physical condition is dominant in each region. Thus, for a given test result, the

CH4	H2	C2H6	C2H2		CO2	CO	C2H4
ELECTRICAL DISCHARGE				SEVERE HEATING			
.	←		LOCAL HEATING	→			

FIGURE 20.1 Conventionally monitored gases and causes for their production (Elzagzoug, 2013).

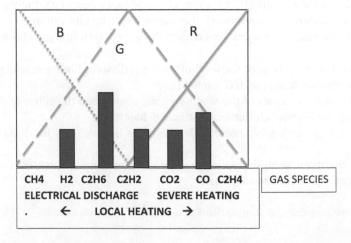

FIGURE 20.2 Chromatic processors (R, G, B) addressing the various gas groups (Elzagzoug et al., 2014).

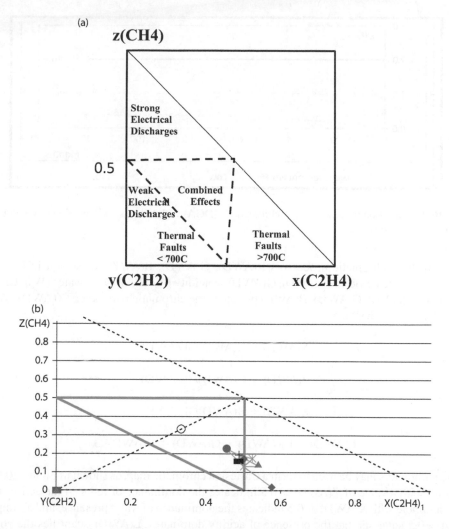

FIGURE 20.3 X, Y, Z Cartesian chromatic map of dissolved gases: (a) different gases and their causes, (b) time variation of various gases [Z(CH4)→strong electrical discharging; X(C2H4)→severe heating; Y(C2H2)→low heating, weak electrical discharging].

relative dominance of each physical condition is conveniently displayed by the location of the test result on this chromatic map (Figure 20.3b) (Elzagzoug et al., 2014). As a result, the variation in the condition of an oil sample with time may then be displayed as a locus of points on the X(DGA), Y(DGA), Z(DGA) chromatic map (Figure 20.3b).

In practice, a knowledge of the effective magnitude L(DGA) of any condition changes is essential for providing an indication of the level of concern which requires consideration. Such an indication may be provided by tracking the change of the effective magnitude L(DGA) as a function of time, as shown in Figure 20.4. Critical values of L(DGA) which indicate normal or abnormal oil conditions are obtained with respect to international scales recognised by the international regulatory body CIGRE

20.3 PRIMARY CHROMATIC ANALYSIS OF ACIDITY, WATER AND DIELECTRIC STRENGTH

The outputs from tests for water, acidity and dielectric (AWD) strength (Shoureshi et al., 2004) form three discrete signals which can be chromatically processed by regarding each output as

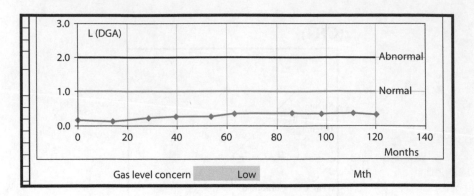

FIGURE 20.4   Variation of dissolved gas chromatic L (DGA) with time as an indication of CIGRE normal and abnormal oil condition.

values from three chromatic processors R(AWD), G(AWD), B(AWD), as shown in Figure 20.5 [i.e., B(AWD) = dielectric strength (D), G(AWD) = acidity (A), R(AWD) = water (W)]. Chromatic processing of R(AWD), G(AWD), B(AWD) produces the chromatic parameters X(AWD), Y(AWD), Z(AWD) defined in Chapter 1.

$$X(AWD) = R(AWD)/3L(AWD)$$

$$Y(AWD) = G(AWD)/3L(AWD)$$

$$Z(AWD) = B(AWD)/3L(AWD)$$

$$L(AWD) = [(R(AWD) + G(AWD) + B(AWD)]/3$$

These parameters may be used to form a Cartesian chromatic map, as shown in Figure 20.6. Thus, a test result with a value of Z(AWD) > 0.5 indicates the dominance of reduced dielectric strength, whereas a sample with X(AWD) > 0.5 indicates the dominance of water presence, and a sample with Y(AWD) > 0.5 suggests that the presence of acidity dominates. L(AWD) quantifies the combined effective strength of the reduced dielectric strength, acidity and water dilution.

Thus, the chromatic map indicates the relative contributions from these three defects (Elzagzoug et al., 2014).

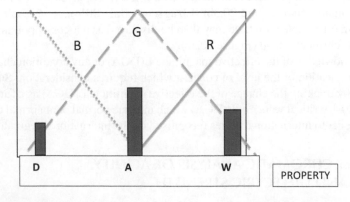

FIGURE 20.5   Chromatically addressed three oil properties B = dielectric strength (D) G = acidity (A) R = water (W) (Elzagzoug et al., 2014).

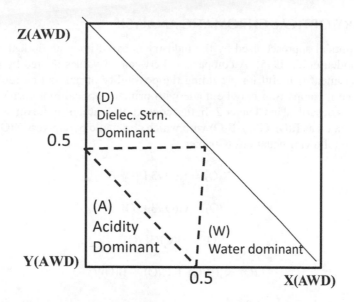

FIGURE 20.6 X, Y, Z chromatic map with interpretation of physical feature acidity (A)→Y(AWD) = A/3L, water content (W)→X(AWD) = W/3L, dielectric strength (D)→Z(AWD)/3L; effective magnitude L(AWD) = (A + W + D)/3 (Elzagzoug et al., 2014).

Recourse needs to be made to the value of L(AWD) in order to obtain an indication of the severity of the deficiencies. Hence, a primary indication of degradation which may occur with time may be obtained with a graph of L(AWD) versus time. An example of such a graph is given in Figure 20.7, which shows the time variation of the level of degradation L(AWD) of an oil sample over a period of several months. Three sectors corresponding to different levels of L(AWD) severity are indicated corresponding to normal, abnormal and intermediate degradation levels. The division between the three levels is based upon the degradation estimates made according to the international regulatory body, CIGRE (Leibfried et al., 2002). The results show the oil to have been close to the lower end of the intermediate degradation zone until month 8, followed by an increase into the uncertain zone at month 10. The oil failed between months 12 and 14, was reconditioned and became approximately normal by month 15.

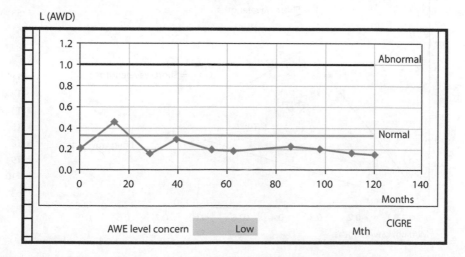

FIGURE 20.7 Chromatic calibration graph for variation of combined AWD effective magnitude L(AWD) with time (normal and abnormal oil condition as defined according to CIGRE).

20.4 PRIMARY OPTICAL CHROMATIC ANALYSIS

A conventional optical approach used by the industry is based upon an optical parameter called colour index (Golden et al., 1995). A comparison between CI values quoted by the industry and a chromatic assessment is useful for providing the possibility of gaining more detailed physical implications. Such a comparison may be made with primary optical monitoring based upon the chromatic methods described in Chapter 2. In this context, the outputs of the three optical detectors R, G, B are designated as R(O), G(O), B(O) with which chromatic parameters X(O), Y(O), Z(O) are determined by the following equations (Chapter 1)

$$X(O) = R(O)/3\,L(O)$$

$$Y(O) = G(O)/3\,L(O)$$

$$Z(O) = B(O)/3L(O)$$

$$L(O) = [(R(O) + G(O) + B(O)]/3$$

These parameters are used to produce an optical chromatic Cartesian map (Chapter 2), an example of which is given in Figure 20.8. This shows several oil samples with different levels of degradation distributed throughout the map ranging from normal (T1), with a dominance of short wavelengths [Z(O) > 0.5], to high-degradation ones, with a dominance of long wavelengths (T9, T10, 63, 85).

The results suggest that a useful parameter for indicating degradation extent may be based upon a shift of a point from short to long wavelengths. This may be indicated in principle via the gradient of the locus from the X(O) = Z(O) = 0 point to the test result point that is X(O)/Z(O).

In order to normalise the X(O)/Z(O) range from 0 (normal oil, T1) to unity (heavily degraded oil, 85) the following formulation has been used:

$$Fn(X(O)/Z(O)) = (1/2)(X(O)/Z(O)) \quad \text{for } X(O) < Z(O)$$
$$= 1 - (1/2)(Z(O)/X(O)) \quad \text{for } X(O) > Z(O)$$

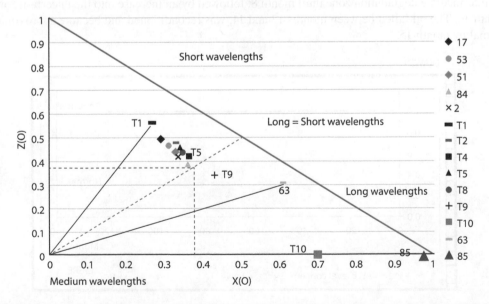

FIGURE 20.8 X(O), Y(O), Z(O) chromatic map for various in-service transformer oil samples (Elzagzoug, 2013). [Sample T1 Z(O)/X(O) > 1; Sample 63 Z(O)/X(O) < 1.]

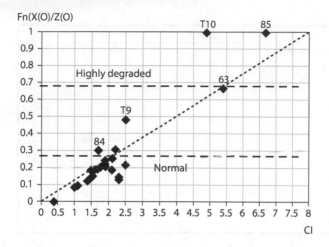

FIGURE 20.9 Optical chromatic parameter [Fn(X(O)/Z(O))] for variously degraded oil samples (CIGRE) compared with colour index (CI) (Elzagzoug, 2013). [For X(O) > Z (O) Fn[X(O)/Z(O)] = (1/2) (X(O)/Z(O)); For X(O) < Z(O) Fn[X(O)/Z(O)] = 1 – (1/2)Z(O)/X(O)].

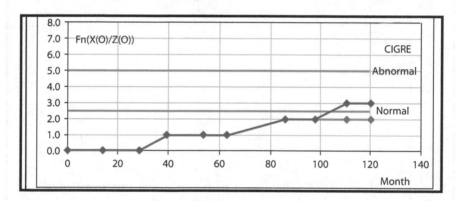

FIGURE 20.10 Variation of colour index magnitude [L(CI) = Fn(X(O)/Z(O))] with time as an indication of CIGRE normal and abnormal oil conditions.

The variation of Fn(X(O)/Z(O)) as a function of the different oils shown in Figure 20.8 is given in Figure 20.9. Results obtained with this chromatically derived parameter are shown to be in good agreement with the colour index results obtained by the industry (Figure 20.9). The results suggest that normal oils lie in the range 0 < Fn(X(O)/Z(O)) < 0.25, whereas heavily degraded oils lie in the range Fn(X(O)/Z(O)) > 0.5 (Elzagzoug et al., 2014).

An example of the time variation of the optical component of an oil is shown in Figure 20.10. Three sectors corresponding to normal, abnormal and intermediate degradations are indicated.

These results show how the colour indication gradually increases with time to be close to the intermediate boundary.

20.5 SECONDARY CHROMATIC PROCESSING

The outcomes from the three primary chromatic processing (dissolved gases [DGA], water acidity, dielectric strength [AWD], colour index [CI]) may be combined by secondary chromatic processing to provide an overall indication (Elzagzoug et al., 2014) of the condition of an oil sample. Secondary chromatic processing involves ascribing scores to each primary chromatically processed group which reflect the degree of degradation based upon empirically determined evaluations as described

(a)

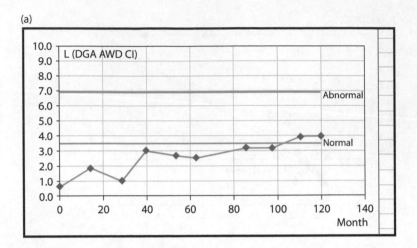

(b)

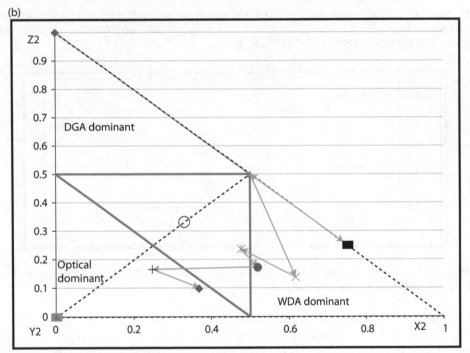

FIGURE 20.11 Secondary chromatic indication of oil degradation: (a) time variation of overall effective magnitude L2 = [L(DGA) + L(AWD) + (CI)]/3, (b) overall chromatic map for the three primary data sets (Z2 = dissolved gases, X2 = water; acidity, dielectric strength, Y2 = optical) (Elzagzoug et al., 2014).

previously. The scores correspond to the values of L(AWD), CI = Fn(X(O)/Z(O), L(DGA)) from the primary chromatic analyses of each set of data, which leads to secondary X2, Y2, Z2 parameters defined by

$$X2 = L(AWD)/3L2 \quad Y2 = (CI)/3L2 \quad Z2 = L(DGA)/3L2$$

$$L2 = [L(DGA) + L(AWD) + (CI)]/3$$

L2 represents the overall effective magnitude of the combined magnitudes of the three sets of primary chromatic contributions of each of the three groups (AWD, CI, DGA)′.

A typical variation with time of L2 is shown in Figure 20.11a, along with empirically determined normal and abnormal boundaries.

The relative contributions to L2 by each of the primary groups (AWD, CI, DGA) may be displayed in a secondary X2, Y2, Z2 Cartesian map, as shown in Figure 20.11b.

20.6 SUMMARY

The condition of an oil may be addressed by chromatically processing data from conventional measurements of various properties such as dissolved gases, the level of water, acidity, dielectric strength and colour index. Each of these three groups of measurements is chromatically analysed separately to produce primary chromatic parameters X1, Y1, Z1, L1. The effective magnitude of each of the three components then form R, G, B values which are addressed chromatically to form secondary chromatic parameters X2, Y2, Z2, L2. Comparison of the overall effective magnitude (L2) with empirically established critical levels provides an overall indication of whether the oil is unacceptably degraded.

The various chromatic maps so produced may be used to trace the source of any unacceptable condition, as shown in the analysis flow chart shown in Figure 20.12 This indicates the traceability from the overall effective magnitude (L2) via a secondary chromatic map (X2, Y2, Z2) to the dominating primary group (DGA, AWD, CI). The chromatic map of the dominating primary group is then inspected to determine which component (X1, Y1, Z1) dominates, identifying the responsible specific oil property and hence the main cause of the degradation.

The chromatic approach has been shown to statistically perform well in comparison with individual oil property tests and also a conventional dissolved gas approach (Duval; e.g., Wang, 2000). The chromatic approach provides a higher sensitivity (Sn, Appendix 5.1, Chapter 5) than Duval analysis (86.7% compared with 26.7%) and has a higher specificity(Sp, Appendix 5.1, Chapter 5) (97.3% compared with 77.8%)

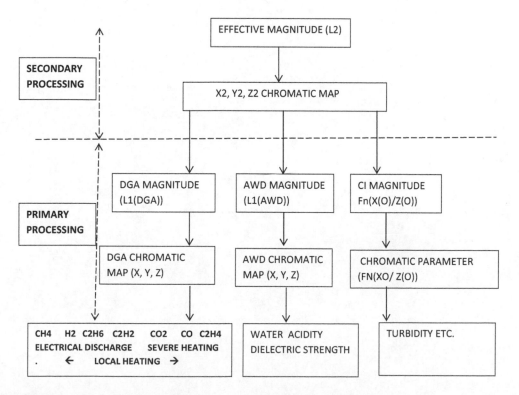

FIGURE 20.12   Flow chart for primary and secondary chromatic processing.

REFERENCES

Aragon-Patil, J., Fischer, M., and Tenbohlen, S. (2007). Improvement of interpretation of dissolved gas analysis for power transformers. *International Conference on Advances in Processing, Testing and Application of Dielectric Materials*. Oficyna Wydawn, Politechn, 2009.

Chiesa, N. (2010). Power Transformer modeling for inrush current calculation. Ph. D. thesis, Norwegian University of Science and Technology, ISBN 978-82-471-2086-6.

Elzagzoug, E. (2013). Chromatic monitoring of transformer oil condition using CCD camera technology. Ph.D. Thesis, University of Liverpool, UK.

Elzagzoug, E., Jones, G. R., Deakin, A. G. and Spencer, J. W. (2014). Condition monitoring of high voltage transformer oils using optical chromaticity. *Meas. Sci. Technol.* 25, 065205.

Golden, S., Craft, S., and Villalanti, D. (1995). Refinery analytical techniques optimize unit performance. *Hydrocarbon Processing*, 74, 85–106.

Jones, G. R., Deakin, A. G. and Spencer, J. W. (2008). *Chromatic Monitoring of Complex Conditions*. CRC Press, ISBN 978-1-58488-988-5.

Leibfried, T., Kachler, A., Zaengl, W., Der Houhanessian, V., Kchler, A. and Breiten-bauch, B. (2002). Ageing and moisture analysis of power transformer insulation systems. *CIGRE Session*, Paris, 1–6.

Looe, H.M. and Zulfadhly, Z. (2015). Condition assessment of power transformer with smart processing of a combination of measured parameters. Asset Management Conference of Electric Utilities (AMCEU), Malaysia, KL, August 17–19.

Shoureshi, R., Norick, T. and Swartzendruber, R. (2004). *Intelligent transformer monitoring system utilizing neuro-fuzzy technique approach*. Intelligent Substation Final Project Report, 4–26. Colorado School of Mines.

Wang, Z. (2000). Artificial intelligence applications in the diagnosis of power trans-former incipient faults. Ph.D Thesis, University of Liverpool.

21 Chromatic Assessment of High-Voltage Circuit Breaker Gases

G. R. Jones, J. W. Spencer, L. M. Shpanin and J. D. Yan

CONTENTS

21.1 INTRODUCTION TO CURRENT INTERRUPTION

Switching electric currents in high-voltage circuits involves mechanically separating two contacts to form an electric arc plasma which is then quenched. This operation is conducted in a tank containing a vacuum or one type of gas which has good arc-quenching properties. The quenching process may be enhanced by blowing the gas through the plasma or electromagnetically driving the arc through the gas.

Monitoring such circuit breakers is important to ensure that there are no power supply disruptions. The monitoring involves addressing many parameters such as the current flowing, various voltages, gas pressure, movement of the switch contacts and so on. Interpreting the monitored results can be a complex process in which chromatic processing can assist. Examples are given comparing data from a small group of medium-voltage air circuit breakers with electromagnetic arc control and non-B field interrupters with several different arc-quenching gases both experimentally and theoretically.

21.2 CHROMATIC INTERPRETATION OF CONVENTIONALLY MEASURED CIRCUIT BREAKER PARAMETERS

21.2.1 INTRODUCTION TO CURRENT INTERRUPTION PARAMETERS

As a result of controlling and quenching an arc plasma during the interruption of high alternating currents, there are complicated variations in electrical parameters of the switch such as the high

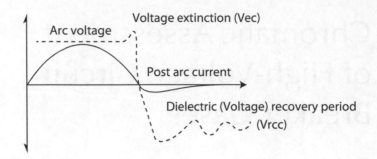

FIGURE 21.1   Parameters evaluated from the tests.

current phase arc voltage peak (Vpkc), arc extinction voltage peak (Vec), peak recovery voltage [V(rcc)] and peak fault current (Ipkc) (Figure 21.1) (Jones et al., 2008; Shpanin et al., 2009). The current interruption process will also be influenced by additional features such as the pressure of the gas inside the switch (pc), the gap length between the contacts (gc) and so on (Jones et al., 2008; Shpanin et al., 2009). In one form of switch, the electric arc may also be controlled by a separate magnetic field (Bc) (Shpanin, 2006; Jones et al., 2008; Shpanin et al., 2009). There is therefore a need for an approach to extract relevant information from such a complexity of data in a manner which is traceable and capable of identifying emerging conditions. This may be achieved through the use of chromatic processing of the data.

21.2.2   Chromatic Analysis Method

Applying the chromatic approach to the current interruption situation involves first ordering and classifying the operational parameters of the circuit breaker in a suitably normalised form into three groups (Shpanin, 2006; Shpanin et al., 2009) (Appendix 21A).

Group I includes *a priori*-determined control parameters (gas pressure, contact gap, arc controlling magnetic field [B]), group II features the proposed interruption conditions (B field, fault current [Ipk], arc voltage [Vpk]) and group III provides the interrupter responses (Vpk, extinction peak [Vext], recovery voltage [Vrc]). The B field belongs to both groups I and II, while the arc voltage belongs to both groups II and III. The normalisation is arranged so that an increase in a parameter value represents a reduction in interrupter performance. Three overlapping chromatic processors (R, G, B), corresponding to each group (I, II, III) are superimposed upon the ordered set of parameters (Figure 21.2a). The outputs from each processor (R, G, B) are fed into chromatic algorithms (Figure 21.2b) to yield three chromatic parameters H, S, L (Chapter 1). Polar chromatic maps of H versus L and H versus S may then be produced to indicate the switching condition (Figure 21.2c).

Values of the H parameter indicate the dominance of each of the seven interruption parameters (gc through to Vrec). Values of L indicate the severity of conditions, and values of S indicate the spread of the influence among various parameters.

21.2.3   Examples of Chromatic Analysis Application

Figure 21.3a shows the normalised values of the operational parameters (gc, pc, Bc, Ipkc, Vpkc, Vec, Vrc) for an air-filled, magnetic field–driven arc interrupter (Ipk = 13.3 kA, Bpeak = 345.3 mT, g = 0.10 6 m) with R, G, B processors superimposed. An H–L polar diagram is shown in Figure 21.3b with results for this interrupter at different peak currents lying in the range 0 < H < 56. This implies that the interrupter's operational conditions are dominated by gap length and gas pressure.

Figure 21.3b shows results for other interrupters – non-rotary arcs in nitrogen air and SF$_6$, air interrupters with B field producing coils. These results show

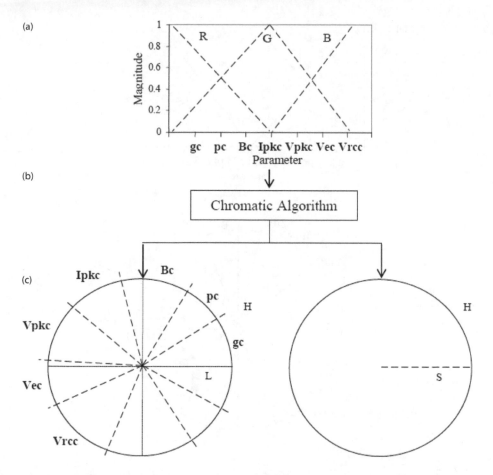

FIGURE 21.2 Chromatic analysis of current interrupter data. (a) Non-orthogonal processors R, G, B superimposed on the three groups monitoring data; (b) chromatic transformation R, G, B>H:L, H:S; (c) H:L; H:S polar diagrams indicating location of components of the three monitoring groups.

1. For the non-B field interrupters, the 1 bar air-filled interrupter had $L \sim 0.7$, the 3 bar N2 interrupter had $L \sim 0.65$ and the SF6 interrupter had $L \sim 0.55$.
2. For the B field interrupter with 1 bar air, $L = 0.65 \rightarrow 0.75$.

Figure 21.3c shows a more expanded H–S Cartesian chromatic map. This shows that for all of the interrupters and conditions investigated, no single parameter is outstandingly dominant, since in all cases, $S < 0.34$ ($S = 0$ – equal influence; $S = 1$ – single totally dominant feature). The variation with peak current is indicated by arrows.

For an air-filled, magnetic field–driven arc interrupter, Figure 21.3b and c show that increasing the contact length further equilibrates the effects of the various parameters (S reduced from 0.22 to 0.12) by promoting the influence of Bc, pc($H \rightarrow 60$) while reducing the stress level (L reduced from 0.725 to 0.64). (Further details may be found in Shpanin et al., 2009.)

21.2.4 Summary and Overview

The application of chromatic analysis enables chromatic maps to be produced which clearly and simply highlight the relative significance of various operational parameters for different types of current interrupters. The parameters involved are divided into three groups corresponding to control

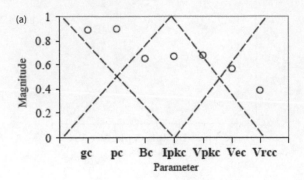

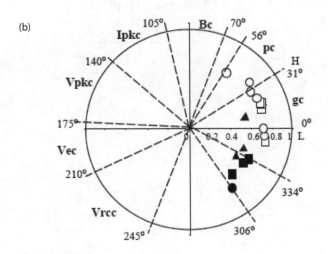

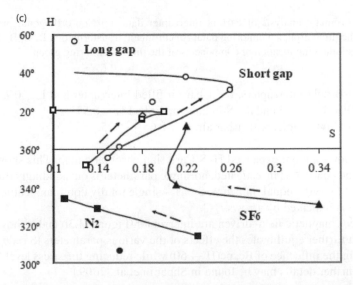

FIGURE 21.3 Chromatic processing of data for current interrupters. (a) Data set ordering with non-orthogonal processors superimposed (example data for B field air interrupter with I peak = 13.3 kA, B peak = 345.3 mT, gap = 0.106 m); (b) polar H:L diagram; (c) Cartesian H:S diagram (expanded region; arrow indicates increasing fault current). (Interrupters: No B field : "▲" SF6, "■" N2, (3Bar); "●" Air (1 Bar), With B field : "◊" "□" "○" Air (1 Bar) (Shpanin, 2006).

parameters, current interruption conditions and interrupter responses so that guidance is provided for possible design improvements.

21.3 COMPARISON OF HIGH-VOLTAGE ARC-QUENCHING GASES: TEST DATA

21.3.1 INTRODUCTION TO ARC-QUENCHING TEST DATA

A widely used gas for high-voltage circuit breakers is sulphur hexafluoride (SF6) (Glaubitz et al. 2014) However, SF_6 also has a high global warming potential (GWP $\sim$ 23500) over a long time horizon (100 years) (Stocker et al., 2013). Its use is therefore regulated and restricted. A search for alternative gases has been ongoing for about two decades (Kieffel et al., 2014; Seeger et al., 2017). The complexity of comparing various alternatives to SF_6 and the amount and variety of information available have contributed to the difficulty of assessments and to find a solution. Such difficulties may be alleviated using chromatic approaches.

Central to the consideration of the choice of gas for arc quenching for high-voltage alternating current (AC) interruption are three of the parameters considered in Section 21.2; see Figure 21.1.

1. The highest peak AC (Ipk) which can be interrupted
2. The dielectric strength (Vd) of the gas
3. Gas pressure change due to decomposition caused by arcing (Dp)

A chromatic-based comparison of the previous parameters can provide a quantitative comparison of various gases.

21.3.2 CHROMATIC ANALYSIS

Potential arc-quenching gases considered by several investigators are carbon dioxide (CO_2) and fluoroketone (FK) plus fluoronitrite (FN), each mixed with air or CO_2 (Uchii et al., 2002, 2004; Kieffel et al., 2016; Seeger et al., 2017). Values of Ipk, Vd and Dp for these gases may be normalised respectively with regard to their values for SF_6, (FN + air) and SF_6. The three normalised parameters are treated as the outputs from three chromatic processors, that is, R = (Ipk)$'$=Ipk/Ipk(SF_6), G = (Vd)$'$ = Vd/Vd(FN + air), B = (Dp)$'$ = Dp/Dp(SF_6).

These can be transformed into chromatic parameters L$'$, X$'$, Y$'$, Z$'$ (Chapter 1)

$$L' = [(Ipk)' + (Vd)' + (Dp)']/3 = \text{effective strength of the three parameters}$$

$$X' = (Ipk)'/3L' = \text{relative magnitude of } (Ipk)'$$

$$Y' = (Vd)'/3L' = \text{relative magnitude of } (Vd)'$$

$$Z' = (Dr)'/3L' = \text{relative magnitude of } (Dp)'.$$

A chromatic parameter $(1 - L')$ may be used to provide an overall quantification of each arc-quenching gas mixture, whilst the X$'$, Y$'$, Z$'$ parameters indicate the relative significance of each component (Ipk)$'$, (Vd)$'$, (Dp)$'$. A graph of $(1 - L')$ versus gas type is presented in Figure 21.4, whilst a chromatic map of X$'$ versus Y$'$ is given in Figure 21.5.

21.3.3 INTERPRETATION OF CHROMATIC MONITORING RESULTS

An approximate indication of the overall behaviour of the various gases is provided by the $(1 - L')$ chromatic parameter (Figure 21.4), whilst a more detailed comparison is provided by the X$'$: Y$'$

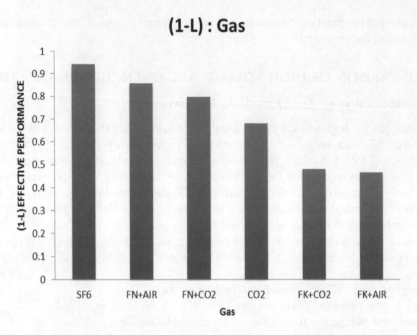

FIGURE 21.4 Effective current interruption performance $(1 - L)$ of various gases $(L = (R + G + B)/3)$.

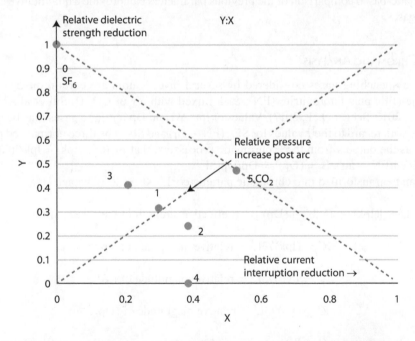

FIGURE 21.5 X : Y : Z chromatic map comparing relative magnitudes of current interruption effects $(0 = SF_6, 1 = (FK + CO_2), 2 = (FK + air), 3 = (FN + CO_2), 4 = (FN + air), 5 = (CO_2)$.

chromatic map (Figure 21.5). The latter emphasises the advantages of SF_6 in having a relatively high dielectric strength $(Y' \rightarrow 1)$, no reduction in the maximum interrupted current $(X' \rightarrow 0)$ and only a low increase in gas pressure $(Z' \rightarrow 0)$.

CO_2 has no post-arc pressure increase $(Z' \rightarrow 0)$, and its current interruption and dielectric strength are lower than SF_6 $(x \rightarrow 0.5, Y \rightarrow 0.5)$. Although the relative current interruption capabilities (X') of the FN and FK gas mixtures are poorer than SF_6, they are better than CO_2. However, the post-current interruption pressure increases (Z′) are greater for all FN and FK gases than both SF_6 and CO_2. Also, the relative dielectric strength reductions (Y′) are all lower than that of SF_6; two of them [(FN + CO_2) and (FK + CO_2)] have similar values to CO_2.

Quantification of the post-arc pressure rise is an indication of the possible production of arc-induced undesirable environmental effects (Chapter 12).

The difference in the values of (1 − L′) for CO_2, FK−$CO_2$0, FN−CO_2 may be ascribed to the complex combination of reduced dielectric strength, current interruption reduction and post-arc production of toxic compounds. Thus, if a slight reduction in dielectric strength is considered less significant for particular applications than a reduction in interrupted current, (FK + CO_2) might be preferable to (FN + CO_2). Consideration of such chromatic factors can assist in decision-making.

21.3.4 SUMMARY AND OVERVIEW

The gas examples treated provide an illustration of how the complex performance of various gases can in principle be conveniently quantified, compared and presented in an X′, Y′, Z′ chromatic map. A calibrated effective performance parameter (1 − L′) can be used for comparing the performance of a gas with a benchmark SF_6. As new SF_6 potential replacement gases become available, the approach can provide a basis for the quantification of performance.

21.4 COMPARISON OF HIGH-VOLTAGE ARC-QUENCHING GASES: THEORETICAL DATA

21.4.1 INTRODUCTION TO ARC-QUENCHING COMPUTATIONS

Considerable computational modelling has been made historically of energy transfer during arc plasma quenching in gas flow, high-voltage circuit breakers (Zhang et al., 2016). Detailed interpretation of such results about the performance of various gases can prove difficult, particularly in identifying any significant property. However, chromatic analysis of such results can assist in the interpretation of such results.

The basis of such computational approaches is a thermodynamic consideration of the balance between the electrical energy input to the arc plasma (V.i); the energy stored in the arc plasma and the energy lost via convection, conduction and radiation. A chromatic examination of these processes for different gases can assist in identifying suitable gases for current interruption.

21.4.2 CHROMATICITY OF THERMAL STORAGE TEMPERATURE VARIATIONS

A comparison may be made of the thermal gas storage (mass density × specific heat) variation with temperature for various gases (Zhang et al., 2016; Guo et al., 2017). The variation of thermal storage with temperature in the range $0–10^4$ K is shown in Figure 21.6 for SF_6, CO_2 and air. The highly complex nature of the storage variations is apparent. The difficulty of interpreting such complex variations may be alleviated by addressing each variation chromatically with three triangular chromatic processors (Figure 21.6), with R covering the electrically non-conducting temperature range 0–5000 K and B the electrically conducting temperature range 5,000–10,000 K.

An X, Y, Z chromatic map for the thermal storage data is shown in Figure 21.7a, with X representing the relative amount of storage for the 0–5,000 K temperature range and Z representing the 5,000–10,000 K range. These results show that at higher temperatures, air has the highest storage, whereas at the lower temperature range, SF_6 has the highest storage. Thus, SF_6 has the advantage of storing a relatively low amount of heat in the electrically conducting plasma range and a considerably higher amount at the non-electrically conducting temperature range.

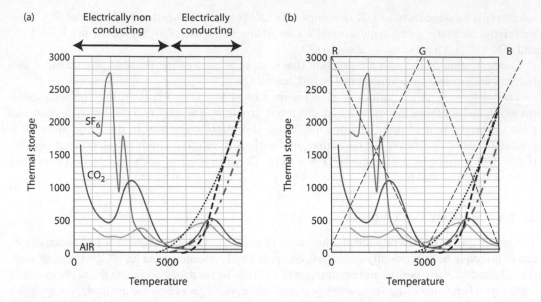

FIGURE 21.6 Chromatically addressed temperature variation of thermal storage of SF_6, CO_2, air. (a) Thermal storage versus temperature; (b) chromatic processors (R, G, B) superimposed upon thermal storage: temperature (CZC paper).

Figure 21.7b shows a graph of the effective magnitude of the heat stored (Ls) versus the spread of the heat stored. SF_6 has the highest stored heat capability with little spread, whereas air has the lowest stored heat with high spread.

These results provide one clear indication as to why SF_6 is a better arc-quenching medium than air or CO_2. It stores only little heat at the electrically conducting temperatures (>5,000 K) but has the highest capability of storing heat at the lower, non-electrically conducting region. Thus, a quest for alternative gases to SF_6 needs to ensure that the relative heat storage between the electrically conducting temperatures is very much less than for the non-electrically conducting temperatures.

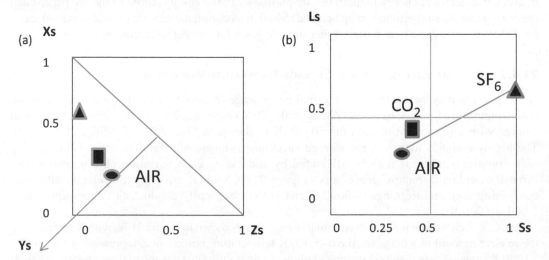

FIGURE 21.7 Chromatic characteristics for thermal storage of electrically conducting and non-conducting temperature ranges for SF_6, CO_2, air. (a) X, Y, Z chromatic map (Xs stored heat, $< 5 \times 10^3$K; Zs stored heat, $> 5 \times 10^3$K). (b) Effective magnitude (Ls) versus spread (Ss) of stored heat.

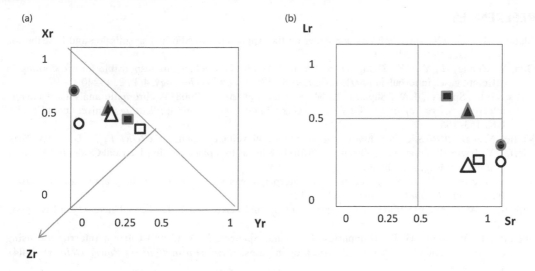

FIGURE 21.8 Chromatic characteristics of radiation, radial conduction and convection of air and SF6 (full symbols – air, open symbols – SF6: circle – 500A, triangle – 50A, square – 25A). (a) X,Y, Z chromatic map (relative values of conduction (Xr), convection (Yr), radiation (Zr)). (b) Effective magnitude (Lr) versus spread (Sr).

21.4.3 CHROMATIC COMPARISON OF DIFFERENT FORMS OF THERMAL LOSSES

A chromatic comparison may be made of the thermal losses from an arc plasma carrying different currents (25–500 A) due to radiation, thermal conduction and convection. In this case, the chromatic processors can be applied so that R=thermal conduction, G=thermal convection, B=radiation. The resulting X, Y, Z chromatic map is shown in Figure 21.8a for air and SF_6 at various currents (25–500 A). At 25 A, this shows little radiation loss compared with conduction and convection for both air and SF_6, whereas at 500 A, SF_6 has a relatively high radial conduction which is greater than air.

A chromatic map of effective strength (Ls) versus spread (Sv) (Figure 21.8b) shows that although SF_6 and air have similar strengths (0.3–0.4) at the higher current (0.6–0.3), SF_6 has a higher strength than air at 25 A.

21.4.4 SUMMARY AND OVERVIEW

Chromatic processing of detailed computational analysis of electric arc plasma behaviour (e.g., Guo et al., 2017) has highlighted features which warrant consideration when seeking possible replacement gases for SF_6. The need for high thermal storage at non-electrically conducting temperatures (<5,000 K) has been indicated. A high thermal power transfer to this area is also highlighted, with both thermal conduction and radial convection having important roles.

APPENDIX 21A: CIRCUIT BREAKER OPERATIONAL PARAMETERS

- I Gas pressure (p) (p/pn = pc; pn = 1 bar);
- I Contact gap length (g) (1 − l/gn = gc; gn = 1 m);
- II, I Magnetic field (B) (1 − B/Bn = Bc; Bn = 1 T);
- II Fault current (Ipk) (Ipk/Ipkn = Ipkc; Ipkn = 20 kA);
- II, III Arc voltage (Vpk) (1 − Vpk/Vpkn = Vpkc; Vpkn = 1 kV);
- III Extinction peak (Vextn) (1 − Vext/Vextn = Vec; Vextn = 1 kV);
- III Recovery voltage (Vrc) (1 − Vrc/Vrcn = Vrcc; Vrcn = 1 kV).

n=normalising parameter value; c=normalised parameter.

REFERENCES

Glaubitz, P. et al. (2014) CIGRE Position Paper on the Application of SF6 in Transmission and Distribution Networks. *Electra*, 34(274).

Guo, Y., Zhang, H., Yao, Y., Zhang, Q., and Yan, J. D. (2017) Mechanisms responsible for arc cooling in different gases in turbulent nozzle flow. *Plasma Physics and Technology*, 4(3), 234–240.

Jones, G. R., Spencer, J. W., Shpanin, L. M., and Deakin, A. G. (2008) A chromatic analysis of current interrupters, *Proc. of the 17th Int. Conf. on Gas Discharges and their Applications*, University of Cardiff, UK, 153–156.

Kieffel, Y. et al. (2014) SF6 alternative development for high voltage switchgear. *CIGRE Paper D1 – 305*, Paris.

Kieffel, Y., Irwin, T., Ponchon, P., Owens, J. (2016) Green gas to replace SF6 in Electrical Grids. *IEEE Power and Energy Magazine*, 14, 32–39.

Seeger, M. et al.(2017) Recent development of alternative gases to SF6 for switching applications. *Plasma Physics and Technology Journal*, 4(1), 8–12.

Shpanin, L. M. (2006) Electromagnetic Control for Current Interruption, PhD Thesis, University of Liverpool, UK.

Shpanin, L. M., Jones G. R., Humphries, J. E., and Spencer, J. W. (2009) Current interruption using electromagnetically convolved electric arcs in gases. *IEEE Transactions on Power Delivery*, 24(4), 1924–1930.

Stocker et al. (2013) REF 2) Climate Change (2013): The Physical Science Basis. Contribution of Working Group 1 to the Fifth Assessment Report of Intergovernmental Panel on Climate Change https://www.ipcc.ch/report/ar5/wgl/

Uchii, T., Hoshina, Y., Mori, T., Kawano, H., Nsakamoto, T., and Mizoguchi, H. (2004). Investigations on DSF6-free gas circuit breaker adopting CO_2 gas as an alternative arc-quenching and insulating medium, *Gaseous dielectrics* (Ed Christophorou et al.). Springer, Chapter X, pp. 205–210. ISBN 978-1-4613-4745-3.

Uchii, T., Shinkai, T., and Suzuki, K. (2002) Thermal interruption capability of carbon dioxide in a puffer circuit breaker utilizing polymer ablation. *IEEE PES T&D Conference*, 6(10).

Zhang Q., Liu J., Yan, J. D., and Fang, M.T.C. (2016) The modelling of an SF6 arc in a supersonic nozzle: II. Current zero behaviour of the nozzle arc. *Journal of Physics D: Applied Physics*, 49, 335501 (15p).

Epilogue

Advances made in the development and application of chromatic techniques have contributed to producing more efficient relevant information and acquisition, addressing the treatment of complex data sets and applying it to a wider range of physical conditions and processes.

It has been shown how the increased availability of new technological hardware has led to the evolution of compact optical monitoring systems. These extend from the original use of optical fibre sensing to the adaptation of miniature and versatile electronic cameras with PCs, leading eventually to mobile phone-based systems and remotely operated wireless systems. All this is attainable in a cost-effective manner.

The further development of chromatic processing has been described for addressing more complex, multi-dimensional signals from three-dimensional images and utilising a combination of visible and infrared laser signals for spatial monitoring.

The chromatic approach has also been deployed for extracting and quantifying information from sets of complicated data such as factors indicating the degradation of high-voltage transformer oil, environmentally hostile features of various high-voltage electrical insulating gases and availability of electric power production from wind generators at different geographical locations.

The versatility of the approach for effective monitoring of faults with cost-effective and convenient-to-use system components has been demonstrated for rail-track fault monitoring, fish behaviour and so on. Further evolution of chromatic monitoring applications remains to be explored as new components are developed and become available.

Index